FOOTPRINTS

IN

THE SAND

by

Virginia Meehan

First printed 2016

Published by Virginia Meehan O'Shannassy:
Virginia@cowirrie.com

Available through Cowirrie:
www.cowirrie.com/footprints

Printed by Lulu:
www.lulu.com

National Library of Australia Cataloguing-in-Publication entry

Creator: Meehan, Virginia, author.

Title: Footprints in the sand / Virginia Meehan.

ISBN: 9780646964836 (paperback)

Notes: Includes index.

Subjects: Meehan family.

       Families--Western Australia--Biography.

       Parent and child--Western Australia--Biography.

       Western Australia--Genealogy.

Contents

Map

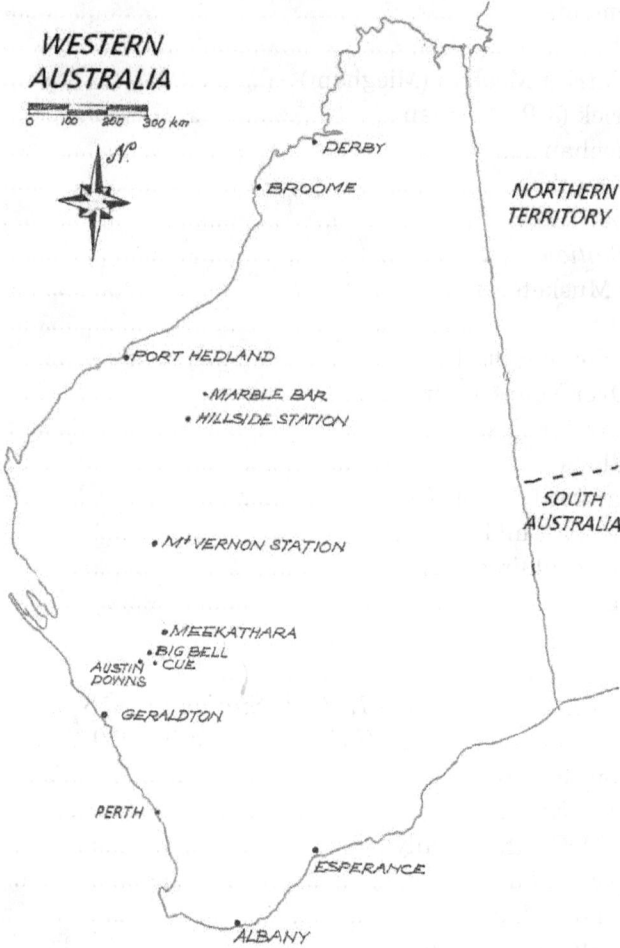

Map © 2016 Frank O'Shannassy

Acknowledgements

Thank you to my sisters, Maureen Stubbs and Elizabeth Krieger; without your support and encouragement. I would never have completed our family's story. Thank goodness for what you could remember. You helped me many times, when I found myself at a loss about events.

Also thank you to my cousins, Barry Meehan, Pam Meehan and Doreen Crispe; you were all instrumental in helping me create this story. You were very generous and giving with the information that you each had about the family. Because of your help I was able to form a sense of who our grandparents and our great grandparents were and also what they had achieved during their lives.

I would also like to thank my cousin Mary (Mem) Partlon who, even though she was very ill in the last few months of her life, went out of her way to give me help, support and encouragement while I was writing this story.

Bree Krieger especially has given me photos. Thanks, Bree, that you're willing to share your photos of the Pilbara. It means so much to me. I think that you will be very happy with the finished story. Thanks Bree and Mike.

Our very close friends Annie and Alf live in Queensland but have been helping me to not give up from the very first. Sadly, Alf passed away on the 6th of July 2016, a few months before I completed the book, but they both deserve a special thanks for the effort they have both made in keeping me at the task. Many thanks.

Thank you to my husband, Frank O'Shannassy, who showed great patience, teaching me to use the computer. He gave me total support and encouragement, and is now doing the research into my family history.

A special thank you must go to our parents, Arthur and Catherine (Kit) Meehan, for without them there would never have been a story to tell.

To my daughter Marita Soutar thank you for not letting me give up even after the computer lost the first draft. Marita, I hope this story will make you very proud of our history.

Thank you to my good friend Jan Horton, who gave so much support and encouragement, even offering to be my willing and patient proof reader, a very big thank you. Without you, Jan, I would have given up but you were always standing behind me ready to pick me up or even catch me when I fell and never once did you lose the whole picture. Thank you so much. I love what we have created. It works so well. I am proud of what it has become.

I so hope that my family will enjoy this story as I have as I wrote about their lives and the disasters that befell those characters during their lives. We have an amazing story filled with love and tragedy.

.oOo.

Introduction

When I first thought about writing about my family; it was to be about my father and mother, Arthur and Catherine (Kit) Meehan, as well as my sisters Maureen, Judith and Elizabeth and growing up with me on cattle stations: *Mt Vernon* in the Murchison, and later at *Hillside Station* in the Pilbara. I wanted to tell our children and grandchildren what life was like for us as children and to give them a sense of belonging to the red dust of the Pilbara and at the same time let them learn about our family history.

These are the footprints in the sands of my memory, like the fragile footprints we were taught to track by our Aboriginal friends back in the Murchison.

As our parents Arthur and Kit Meehan are no longer here to entertain with us with stories about the family, I hold the hope that our children will one day, like us, want to hear more about our family. I also wanted our children and grandchildren to appreciate and understand what our families achieved during their life-times.

Memories can do strange things at times; one minute they are as elusive as butterflies and the next they are like a giant rock weighing you down. I have always had a curiosity about our family's history – where our father's and mother's families came from in Ireland and what made them settle in Australia – in the back of my mind. I kept this thought that one day I would try and tell their story. Unfortunately I think I may have left my run a bit late, as family members from my mother's side (the Murphys) have all passed away. I have been trying to trace their descendants but it is a slow and often frustrating task.

I have been a lot more successful with my father's family, the Meehans. Again I have been trying to trace any Meehan descendants who may have information that I can use. It is starting to look like my family might be the last of our tribe, as there may no longer be any Murphys or Meehans alive to whom I may turn for information on our history.

I know from both our father and mother that our roots are in Ireland; both our parents' families come from Ireland. Great Grandfather William Meehan came from Kilkenny in Ireland. He lived in America for a number of years before he migrated to Australia sometime in the 1850s. He then settled in Australia in the state of Victoria, married and raised his family. Great grandmother Julia came from Mitchelstown in Ireland and her mother was Julia O'Brien and her father was James O'Brien.

William and Julia Meehan are our great grandparents. Their eldest son John (J.P.) Patrick and his wife Ida (Minnie) May Blanche McCarthy Meehan are our grandparents. My father was James Arthur (known as Arthur) Meehan and Catherine (Kit) Murphy Meehan was our mother.

I found that as I wrote more and more about my family, I was learning as well. Learning about the Meehan clan, and our Irish connection due to our grandfather William Murphy and our grandmother Catherine Meagher Murphy. They gave us our Irish background. It has turned out to be quite a journey and I hope the family will enjoy and learn from what I have written.

My wish is that my daughter Marita Soutar and her daughter Georgia Grace and stepsons Alistair and Ian, as well as my sister Maureen's children and grandchildren Julie, Deanne, Janine, Daniel and Shannon as well as Janine's sons, Sebastian, Finn and Miles, and Daniel's daughter Zoe, will learn more about their family from reading this story. I also hope Judy's children and grandchildren Marisa, Matthew, Geordie and Kalan and Liz's children Tye and Bree will enjoy reading and learning about our family. My sisters Maureen Stubbs and Elizabeth Krieger have been great but time has made our memories of that period in our lives somewhat scratchy. The following story will be my memories of that time and of the events that took place, as well as what I was able to find out when doing research about our grandparents and great grandparents as well.

I hope everyone enjoys this journey, back to times that are now long gone.

Permissions

I have made an effort to keep my family up to date with any research that I have found and made sure that I have had my sisters' and cousins' permissions to include all that they have shared.

.oOo.

CHAPTER 1
William Patrick Meehan (Miegham)

William Meehan (Miegham)

Our great grandfather William Patrick Meehan (Miegham) was born in Kilkenny, Ireland. William's mother was Ellen Toohy, and her family were also from Kilkenny in Ireland. His father was John Miegham who was a farmer in Kilkenny.

William changed the spelling of his surname from Miegham to Meehan in the late 1800s; our father told us about this when we were growing up. As far as our father knew this was done because of a falling out between the brothers. From what our father said, a couple of the brothers followed William out to Australia, which ended in family fights. Our father never told us what the fights were about. I don't think our grandfather John Patrick Meehan ever told his family what the problems were.

William's eldest son (our grandfather) John Patrick was born in Australia, but the family still gave their Irish address – 27 Gears Street Kilkenny – on his birth certificate[1].

During this time (1845-1853) Ireland was suffering from the Potato Famine. William's family – like many families in Ireland – had to make the hard decision to let their sons go to Australia or (as in the case of my grandfather and his brother) to America. There they were able to find work and send money home for the family. William was very aware of the toll this would have on his family back in Ireland as he and his brothers were young men at the time.

William Patrick Meehan (Miegham) immigrated to America during the American gold rush, possible dates 1848-1849. He

---

[1] As the writing on the birth certificate was very difficult to decipher I have left a question mark over that information for now.

settled in California from what our father told us. William's brothers followed him out to America some time later. The three brothers were in America for some years. They then left America and came to Australia; that was 1854. At this stage we do not know more about that side of the Meehan family, which is a pity.

I can remember Mother telling me during a phone call home to Perth (I was living in Victoria at that time) that one of the nurses taking care of her while she was in the Royal Perth Hospital (where she was being treated after a house fire) was from America and she had the same surname as ours – Meehan. This nurse checked with her family back in America and it turned out that she was a relative of our father; a distant cousin I believe. This was during the early 1970s, from memory.

Sadly our father and his brother Jack had passed away when we came across this information. Jack passed away in 1962 and our father in 1973, so we were not able to find out any more about this possible family connection. Our Aunty Grace (Father's sister) was not well during this time and so I believe no more was done to bring the families together.

From what I have learnt, Great Grandfather William, with members of his family, became active in the American Wars of that period (Mexican-American War 1846-1848).

William Meehan (Miegham) may have been connected to the Irish Brigade or the Fenian Brotherhood, and it is believed that he may have joined the Union as well. Many Irishmen living in America did so at that time. There was also an Irish-American party, which had the interesting name of the Know-Nothing Party. The Irish were involved in the formation of this organisation, or so the story went.

William with some fellow Californians followed the gold rush from America to Victoria in Australia (possible date was 1854). He spent time in the Ballarat and Kilmore areas.

I did find a mention of a William Meehan (Miegham) from Kilkenny migrating to America and then sailing from America to Auckland in New Zealand before coming to Australia. I am unable at this stage to confirm this information.

Peter Lalor

After arriving in Victoria, William made his way to Ballarat in search of gold. There he met and became friends with Peter Lalor. Below I have included some information about Peter Lalor as well as the relationship that he had with our great grandfather.[2]

Our great-grandfather William also took an interest in the Eureka Stockade and, although he did not become as active in the movement as Peter Lalor, William made it known that he supported Peter and the miners' stand and would help and assist the men who were involved in the Eureka Stockade by providing them with whatever they needed in the way of goods and services. William Meehan believed strongly in their cause and went on to help Peter Lalor and others establish the "Liberty of the Citizens of Australia Movement".

William and Peter Lalor are said to have remained in touch over the years. Our grandfather J.P. Meehan is believed to have met him just after he had left school when his father sent him

---

[2] Peter Lalor was originally from Tenakill in County Laois in Ireland. Peter Lalor's father, Patrick Lalor, was a member of the British House of Commons, and Peter would go on to become a member of the Victorian Parliament in 1856. He was the author of the Oath of Allegiance used by the miners at the Eureka Stockade. "We swear by the Southern Cross to stand truly by each other to defend our rights and liberties".

Peter Lalor became the leader of the Eureka Stockade in about 1854. During the uprising in 1854 many miners were killed by Her Majesty's police and soldiers. The miner's under Peter's guidance stood firm, as it was always Peter Lalor's belief that blood had to be spilled before political reform could take place. "Peace hath her Victories no less renowned than War." Lalor went on to become Speaker in the Victorian Parliament : From Australian Dictionary of Biography, http://adb.anu.edu.au/biography/lalor-peter-3980

down about some business. Peter Lalor and our great grandfather William both died in 1889.

During my research on William Meehan, I came across a James Meehan from Offaly in Ireland. James Meehan it turns out was also in Ballarat in 1854, the same time as our great grandfather William Meehan. I have not been able to find any other connection between the two men but it is quite possible that they were brothers or cousins and James may have been with William in America before following him to Australia[3].

William and Julia

William married Julia O'Brien, a servant girl, the daughter of James O'Brien, a baker and Julia O'Sullivan. They had also come from Ireland and gave their address as 120 Grant Street, Mitchelstown Ireland, in the County of Tipperary.

William and Julia were married in Melbourne Victoria on the 5th of April 1858. Julia was 27 years old and William was 30 years old. The family story was that Julia and William met when they were on the ship from America to Australia and the ship spent some days in New Zealand.

After they were married they lived in the Kilmore area, possibly at Yarck (a small mining town at that time).[4]

William Patrick and Julia had six children: five boys and a girl, born over a span of just seven years. They were John Patrick (Jack or J.P.), Elizabeth Ellen (Bessie), William Francis (Will), James Joseph (Jim), Michael (Mick) and Patrick.

---

[3] 'This information came from a Limerick-based Historian, Ruan O'Donnell. "Irish in the land of Oz", page 7

[4] I have heard it said by members of my family that William Patrick and Julia Meehan were neighbours of the Ned Kelly family. He was the notorious Victorian bushranger. Ned Kelly and his gang have been written up in Australia's history books.

It has been said that John Patrick Meehan (our grandfather) was born in 1859 at Northcote in Melbourne, but I have a copy of his birth certificate which gives his date of birth as the 27th of January 1858. It also gave his place of birth as Heathcote, Victoria. His birth was also registered in Heathcote on the 18th of March 1858; there appear to be some discrepancies in this information.

This was possibly due to the times – 1800s birth and death records were not always recorded accurately, or even submitted by the immediate family.

Figure 1: Julia Meehan

Figure 2: William Patrick Meehan

In 1874, at the age of 16, the eldest son (our Grandfather J.P.) left home to go droving in Queensland. Patrick, the youngest son, was only 9 years old then, and J.P. did not see him again as he died on the 25th of January 1887, at Yarck in the Shire of Alexander, County of Anglesey in Victoria. Patrick is buried in Yarck Cemetery. He was only 22 years, 11 months and 25 days old when he passed away from hepatic abscess (abscesses on the liver) and something called Exhaustion from Phthisis (a progressive wasting disease).

Our great grandfather (William Meehan) was well known in Victoria as an agriculturist. He was a very public spirited man from all accounts and became involved in many community matters. I have also found a mention that he was a member of the police force for some time.[5]

William's sons Will and Jim also left home to join J.P. in droving up north, while Great-Grandfather seems to have got the gold itch again with the Coolgardie gold rush of 1893 and went west after Patrick died.

When I started researching our family looking for any information, like birth and death records, my sister Maureen was kind enough to pass on to me William's death certificate which shows that our great grandfather William Patrick Meehan died on the 6th of July 1895. He died from stomach cancer aged 67 years and was buried in Northam, Western Australia on the 13th July 1895.

William had been in Western Australia, perhaps seeking gold. He became ill in Northam (which is on the road to the Coolgardie gold fields from Perth) and, so the story goes, he was cared for by the widowed Clementina McCarthy and her family until he had to be moved to the Northam Hospital where he died. Clementina's daughter Minnie, who later became J.P.'s wife, was then 16 years old. Perhaps that is how she came to meet J.P.?

My sister Maureen has also given me documents that our father gave to her which included a copy of William Meehan's will, dated the 6th of July 1895 (a few days before he died). William

---

[5] *Alexandra and Yea Standard*, 9 January 1942, page 3. Death of Mr W.F. Meehan.

left all his worldly possessions to his youngest surviving son, Michael. He did not leave anything to his widow, Julia.

Great Grandmother Julia remained at Yarck when William went west, but lost touch with him as he moved about in Western Australia. Eventually she asked the police in Western Australia to find him for her, which they did, but sadly it was too late. William had already died at Northam.

Julia was left living alone in Yarck with no means of support, so she sailed to the North West herself where her sons were settled.

Julia Meehan: her daughter and sons

Our Great-Grandmother Julia had 4 surviving sons and one daughter when she decided to leave Yarck in the 1890s.

Elizabeth Meehan

Bessie Minogue was J.P.'s sister and our great aunt though we never met her. She had married Denis Minogue who was a warder at Geelong Gaol. They had five children: Ethel, John (Jack), Frank, William and Mary. However, baby Mary died at 10 months old, and William died in World War I of illness on his way to the Front in 1918. Then the next year, in 1919, Jack died after his legs were severed in a terrible railway shunting accident at Geelong.

John Patrick Meehan

John Patrick (J.P.) Meehan (our grandfather) married Minnie (Ida May Blanche) McCarthy and developed *Austin Downs Station*, out of Cue. They had three children: Arthur (our father) Uncle Jack and Aunty Grace.

William Francis Meehan

Will Meehan married Helen (Ellen) O'Rourke[6] of Eildon in Victoria. He sold up his holdings at Yarck and went to WA in 1893 to raise his family near Geraldton. They eventually retired back to *Glenira* at Cathkin in Victoria. He died there on the 31st of December 1942 and was buried at Yarck Cemetery with his brother Patrick[7].

Will left a son, John Patrick (Jack) Meehan, and daughters Ellen (Eileen Veronica) Clemann, Mary Meehan and Olivia Falconer – they were our father's first cousins and I got to know the girls well later on. Their brother Jack Meehan had died at Cessnock in 1951[8] so I never met him.

James Joseph Meehan

Jim Meehan[9] was in WA in 1902 working at the *Star of the East* gold mine, and then bought part of *Mt Erin Station* near Geraldton[10] in 1905. He married Annie Ethel Alston of Alexandra (Victoria).

Annie was 29 years old and, with her brother Harry, was visiting their Aunt Ellen Meehan (Will's wife) in Geraldton when she got engaged to Jim. They returned to Victoria to marry on the 12th of March 1906 at St Patrick's Cathedral in Melbourne before returning to Geraldton. Annie was a talented pianist and her brother Harry, who also settled in W.A., played the violin.

---

[6] Michael P. Byrne: Brushy Creek, A Rourke Family History: 3rd Edition, 2016

[7] *The Argus*, 3 January 1942, page 2: Family Notices

[8] *The Argus*, 26 February 1951, page 16: Family Notices

[9] http://rourkefamilyhistory.blogspot.com.au/2016/01/patricks-great-grandchildren.html

[10] *Geraldton Guardian*, 16 June 1908, page 2: Local news

Jim was in partnership at times with his brother J.P in the butcher business at times, including at Nannine and then on his own at Northampton.

Jim and Annie had two sons; Bill (William Patrick) and Jack (John Henry James). They were living at Geraldine where Jim was working the *Surprise Mine* at Ajana[11]. Annie was expecting twins when she became ill with septicaemia. Despite a rushed trip to hospital over difficult roads and in high temperatures, she died on the 2nd of May 1919 aged 42. Annie Alston Meehan was buried in Northampton.[12]

Their uncle J.P. 'kept an eye on the boys', and had his nephews to stay at *Austin Downs* at times I understand while their father slowly recovered from the loss of his wife and twins.

Jim Meehan then took on the management of J.P.'s *Gracemere* properties at Coomberdale. He died in Geraldton on the 16th of June 1948 from bowel cancer. I was only about 7 years old then and don't remember ever meeting him. I do remember Jack and Bill, our father's first cousins, though. Bill took over the management of *Gracemere* from their father, and his brother Jack was there as well for a time. Jack's son Barry has been a wonderful source of information about his family and *Gracemere* for me.

Michael Meehan

Mick Meehan was J.P.'s youngest surviving brother. He does not appear to have married or left any children. He died aged 58 in an accident at Cue on the 7th of September 1929. This was well before our parents' wedding, so we never knew Great Uncle Mick.[13] [14]

---

11 *Geraldton Guardian*, 31 May 1919, page 2: Death at Geraldine

12 *The W.A. Record*, 7 June 1919, page 13: Local and General

13 *The West Australian*, 2 October 1929, page 1: Deaths

14 *The West Australian*, 14 September 1929, page 1: Deaths

Julia's Later Years

Julia's sons took good care of her at the end of her life.

Family stories tell that J.P. and Minnie looked after her at *Austin Downs Station* for some of the time in the five years from their marriage to Julia's death. My father Arthur remembered his Grandmother Meehan there, although he was only about 4 years old when she died. The sad thing was that she had also developed stomach cancer, like their father William.

Figure 3: Michael Meehan

Julia was in Geraldton with her son Will and daughter-in-law Ellen when she died aged 75 on the 30th of July 1907 (twelve years after William's death in Northam).

Our Great Grandmother Julia is buried in what was known at that time as the 'New Cemetery' in Sandy Hollow there in Geraldton. This is now covered over with housing.

.oOo.

CHAPTER 2
John Patrick (J.P.) Meehan

Our grandfather John Patrick Meehan was born in Victoria on the 27th of January 1858, the eldest son of William Patrick and Julia Meehan. He was called Jack when he was young, but was widely known as "J.P." with so many other 'Jacks' in the family.

J.P. grew up in the Kilmore area (grandfather's death certificate shows the spelling of Kilmore as Kilgmore), possibly at Yarck, a country town in Victoria. He left school when he was about 13 or 14 years old, which was quite a common thing for country kids to do at that time. Their parents would then put them to work on the family's farm. It has also been said that J.P. did not like school very much; he preferred to be outside, working on his father's farm or for other farmers in the area.

Grandfather apparently not only disliked school; he also could not stand the teacher, so the story goes. I gathered from Father that the dislike was mutual; the teacher did not like grandfather all that much either. Grandfather was always being told off for playing up in class. In the end it got to a point that he was sent home to be chastised by his parents, which would have been his father by all accounts.

Great-Grandfather was a no-nonsense type of man so, after much discussions, it appears that William and his eldest son decided that it would be better if J.P. left school and went out to work.

The Droving Life

After some time working on the family farm and for neighbours, John Patrick left Victoria and moved to Queensland where he found work as a drover. J.P. would have been about sixteen at this time and settled into the Queensland way of life very quickly. As time went by, he decided that Queensland and the bush offered him a chance at a new life.

Queensland became J.P.'s home for the next 12 years, during which time he worked hard and established a reputation as well as the respect of his peers and pastoralists as a skilled stockman and as a much sought-after drover. John Patrick found himself in great demand by the pastoralists to take their stock to the markets.

Grandfather became familiar with every stock route from the Gulf of Carpentaria in Queensland to Melbourne in Victoria. The time came when John Patrick got bored with what he was doing, so with his younger brother and a friend they decided to set out for Western Australia and the gold rush that was taking place there; that was in early 1889 when he was 31 years old.

J.P. Meehan's Friend: Frederick Vosper

While living in Queensland John Patrick Meehan met and became good friends with Frederick Vosper. In fact, it was because of Frederick Vosper that J.P. decided to move to Western Australia.

Below is a copy of an article written about our grandfather that was published in the *Geraldton Tribune* of 1910. A member of the family came across it while settling a family member's estate.

Meehan, of Austin Downs

A fine robust type of the men who have blazed the pastoral track upon our goldfields is J.P. Meehan of *Austin Downs* Station (right up against Cue and Day Dawn) and a few other holdings in this country of vast spaces.

J.P.M. is a Queenslander, and in that State he was an old friend and staunch admirer of the late F.C.B. Vosper before the many-sided man heard the West-a-calling. But the Murchison squatter of today, the Queenslander stockman and overlander of the eighteen-nineties came West very soon after Vosper and wasn't long in discerning, with his practised eye, the grazing worth of the back country as well as its auriferous attractions. He trans-continented, first in the eighteen-nineties, from Port Douglas to Cue, and dry as the track was in those days, he will still tell the story, of how he carried a bottle of brandy with him (unopened) from one side of Australia to the other.

Twice in the years following he made the same formidable and lengthy trip that the trans-line has now reduced to a pleasant jaunt of a week or so.

In the early days of the Murchison, Mr Meehan busied himself with butchering and stock dealing, but it was not long before he launched out into bigger things.

He took up the *Austin Downs* Station for pastoral purposes, and obtained some of the leases formerly held by Wainwright and Co, the big Geraldton firm, that dabbled in everything till an unkind fate brought them down. Mr Meehan has now been settled on the Murchison for over twenty years, and the best proof of his shrewd judgement and his successful enterprise is that this year he shore 28 thousand sheep. For close on twenty years he has been a magistrate, entitled to wear the same letters before and after his name.[15]

There is nobody who knows the country better and no more enthusiastic believer in it, and no better judge of sheep and cattle than this rigorous and optimistic Murchison pioneer.[16]

After doing some research on Frederick Vosper using the Wikipedia website I was able to find more interesting information about our grand-father's friend. Frederick Vosper immigrated to Queensland in early 1886. He tried droving and mining (which is where he would have met J.P.), and then drifted into journalism before moving to Cue in Western Australia to become the editor of the Murchison Miner in 1893. This is what initially attracted Grandfather J.P. to the area as Vosper was a passionate advocate for new opportunities.

---

[15] My father and mother were very proud that J.P Meehan was a magistrate for twenty years in the Murchison area.

[16] Taken from a copy of the Geraldton Tribune 1910. Also printed in Sunday Times (W.A.) 29 September 1918: Meehan of Austin Downs

Frederick Vosper was elected to the Western Australian Legislative Assembly on the 4th May 1897 but he died suddenly from appendicitis on the 6th of January 1901 aged 31 and is buried at the Karrakatta Cemetery.[17]

Prospecting

J.P., his brother and a friend joined forces with the Duracks who were interested in moving to Western Australia to set up a pastoralist business in the Kimberleys. The Duracks went on to become a part of Western Australian's history and followed the Canning Stock route across from Queensland to Western Australia.

J.P. and his brother drove horses from Georgetown in Queensland to Nullagine in the North West of Western Australia. After delivering the stock, J.P. and his brother set off for the goldfields to make their fortune. From what I have been able to find out, grandfather spent about four years prospecting for gold in Western Australia before he moved on to other ventures.

My search has shown that J.P. spent a great deal of his time in and around Day Dawn, Meekatharra and the surrounding areas with his friends Luke Soich and Tom Porter. They provided the goods and service which the prospectors working in the area needed. J.P. also spent time prospecting around Day Dawn and Cue, as well as Big Bell.

J.P. Meehan, on behalf of his group, in May of 1896 pegged out three goldmining leases in the Meekatharra area. This is an interesting bit of Meekatharra history as those gold leases were the first leases worked in the Meekatharra area. Because of that find, the local Shire at that time named a street in

---

[17] Wikipedia – Frederick Vosper, https://en.wikipedia.org/wiki/Frederick_Vosper, accessed 11Nov16.

Meekatharra 'Meehan Street' after our grandfather J.P. Meehan.

Family gossip has it that the first part of the spelling of Meekatharra incorporated the name Meehan (Mee) as a thank you to J.P. and his friends Luke Soich and Tom Porter[18].

Finally Grandfather became bored with prospecting and decided to open a butcher's shop in Cue. Family history says that when J.P. was excavating the land for his butcher shop, he found gold and this gave him a good start on purchasing *Austin Downs*.

J.P. did go on with the butchering business, running it in conjunction with the station. He had shops at Cue, Day Dawn and Nannine. The Nannine shop was run in partnership with his brother Jim in 1897[19], and Jim also had his own butcher shop at Northampton.

Party at the Day Dawn Hotel

All three of J.P.'s surviving brothers (Will, Jim and Mick) had moved west too. One interesting bit of folk-lore that I came across was that two of J.P.'s brothers bought the Day Dawn Hotel. This would have been in the late 1890s I believe. The story goes that the week that they moved into the hotel they invited all their mates in for a drink. The pub was then closed to everyone else.

During the course of that week the brothers and their friends managed to drink the pub dry. It was some time later that the pub's bar was restocked and opened again to the public, or so I have been told. As most of the brothers' friends, like themselves, hailed from Ireland or were of Irish descent, they knew how to drink and enjoyed a good fight as well. So by all accounts,

---

[18] The History of the North West of Australia. Edited by J.S. Battye. P11

[19] Wikipedia – Nannine
https://en.wikipedia.org/wiki/Nannine,_Western_Australia#Other_Businesses, accessed 11Nov16.

barring hangovers and black eyes, every one claimed that they had had a smashing good time.

Austin Downs

In the mid-1890s I understand grandfather purchased *Austin Downs Station*, a large pastoral property of nearly 400,000 acres which had been founded by the late Lacey Brothers.

In 1908, J.P. was mentioned in 'A Mustering of the Greybeards' of the Murchison where it was said of him:

> Jack Meehan ... possesses interests in solid business propositions and has put a lot of money into mining. He can write the letters J.P. at each end of his name. He is happiest most, in his white moles and cabbage tree hat, when cutting through the mulga to have first say in a deal of 'fats.'[20]

Figure 4: Austin Downs homestead

---

[20] *Murchison Advocate* 25 July 1908, page 3:The Old Brigade

J.P. bought the station partly with his profits from selling the gold leases. He established *Austin Downs* to be his family home and to raise sheep and cattle there. He then sold the beef and lamb in his butcher shops. In 1916 his holding was described as:

> ... he has a fine property in his *Austin Downs* station, near Day Dawn. His holding comprises 396,000 acres, with 48 windmills and 400 miles of fencing erected, the whole being subdivided into 28 paddocks. This year he shore 16,516 sheep, but in 1910 he shore 22,000. Besides the sheep he has 1000 head of cattle and about 90 horses at present running on the property.[21]

Gracemere

In 1910, Grandfather acquired *Gracemere*[22] which he named after his only daughter Grace, then three years old. This property was situated at Coomberdale, just north of Moora and was run as part of 'J.P. Meehan and sons'. *Gracemere* was around 30,000 acres, and included a 4,000 acre freehold farm (called *Gracemere Park*) which was on the eastern side of the Midland Railway line at Coomberdale[23] and thus very convenient for loading and unloading stock.

Grandfather J.P. employed his brother Jim Meehan as manager of the *Gracemere* properties after Jim's wife died. Jim's sons, Jack and Bill, eventually worked there too. Bill in turn became manager until J.P.'s son Jack inherited the 'J.P. Meehan and son' holdings.

---

[21] *Meekatharra Miner*, 24 December 1916

[22] *Northern Times*, 17 January 1947, page 12: Obituary the late Mr J.P. Meehan. A prominent Murchison Pastoralist

[23] *Geraldton Guardian*, 9 August 1924, page 2: Pastoral Notes "Mr Meehan has made arrangements to convey to his Coomberdale property, on the Midland, some 3000 head of sheep, owing to shortage of feed here."

J.P. used this more southern property to breed cattle and sheep. He ran his own stud there, and the young stock were then shipped north to Austin Downs until they were ready for market or breeding. *Gracemere Park* was used for mixed farming; grain crops, sheep, cattle and pigs. J.P. would send his sheep and cattle down to this farm to be rested and finished off before they were then moved on down to the markets in Perth. The accessible railway transport was a vital part of this complex and very practical operation.

Other interests

In 1929, J.P. sent his son Arthur north to find more land, and 'J.P. Meehan and sons' took on *Mt. Vernon Station*. This was intended to become exclusively a cattle property, with the sheep maintained at *Austin Downs*.

It has been brought to my attention by family members that J.P. Meehan – our grandfather – also had property in and around Kalgoorlie as well as Meekatharra. There is also talk about land that he owned in Perth in St Georges Terrace. I have not as yet been able to trace any of these properties so at this stage cannot confirm what I have been told.

Community Affairs

J.P., like his father William before him, was very interested in his community's affairs and became involved in many of the districts activities. These included membership of the Nannine Municipal Council, the Murchison Roads Board and many of Day Dawn's organisations such as the hospital committee. He was a member of the Cue Road Board at various times.[24]

It is my understanding that J.P. Meehan was also involved in setting up the site for the new mining town of Big Bell in 1936. The site was originally part of *Coodardy Station* and one of the

---

[24] The History of the North West of Australia. Edited by J.S. Battye. P11

station's paddocks was excised for this purpose. Grandfather proposed the main street be named Coodardy Street. Meehan Street was proposed by a Mr Hince[25].

Grandfather (from what the family have said) was very proud to have his name used as a street name in the new town of Big Bell. However, times change, and all these years later Big Bell is now a 'ghost town'.

Herbert Hoover

During grandfather's prospecting days he met and became friends with Herbert Clarke Hoover[26], a mining engineer who was working in Western Australia during the gold rush years.

At around this time Grandfather would visit Kalgoorlie regularly to check on properties that he owned in the Kalgoorlie area and would from what we have been told do a spot of prospecting as well while he was there. From all accounts this is where Herbert Hoover and our grandfather met and became close friends. They remained friends and stayed in touch until our grandfather died in 1946.

It may have been on one of his trips through to Kalgoorlie that he stopped at Northam and met his future wife, Minnie McCarthy. She was apparently working as a nurse at the

---

[25] Wikipedia – Big Bell, https://en.wikipedia.org/wiki/Big_Bell,_Western_Australia, accessed 11Nov16.

[26] Herbert Clarke Hoover joined Bewick, Moreing and Co, a British mining and engineering company in 1897. That same year he was sent out to Western Australia, where he worked as an inspecting engineer, evaluating mines and prospects in the Kalgoorlie and Coolgardie area. When the *Sons of Gwalia Mine* opened near Leonora in 1898, Herbert Clarke Hoover became the mine manager. Herbert Clarke Hoover went on to become the 31st President of America (1929-1933); he would only serve one term. From: Wikipedia – Herbert Hoover, https://en.wikipedia.org/wiki/Herbert_Hoover, accessed 11Nov16.

Northam hospital. From what the family have said, it was love at first sight when they met at a social being held at Northam.

Marriage: J.P. Meehan and Minnie McCarthy

J.P. and Ida May Blanche McCarthy (known as Minnie or Min) were married on the 4th of June 1902 at the Catholic Church in Northam, which was her home town.

Minnie was the daughter of Denis McCarthy and Clementina Langoulant of Northam. Her father had died when she was 9 years old and her mother was blind. (See Appendix 1 for her family story).

Denis had been born in Ireland but Clementina was born in Perth to very early settlers: English-born Mary Anne King and the French Louis Langoulant.) See Appendix 2 for the King story and Appendix 3 for the Langoulant story.)

There was quite an age difference between J.P. and Minnie. John Patrick was 44 years old and his bride was only 22 years of age, but by all accounts their marriage was a strong and loving one. Their wedding was described in some detail in the local newspaper.

> A Wedding was celebrated by the Rev. Fr. Walsh in St. Joseph's R.C. Church, Northam, on Wednesday June 4, between Miss Ivy May Blanche McCarthy, of Northam, fourth daughter of Mrs. McCarthy, *Thankful Rest*, Northam, and Mr. J. P. Meehan, J.P. of Day Dawn. There was a very large gathering of friends to witness the Ceremony. The Bride, who was given away by Mr. Bert Donovan, of Kalgoorlie, wore a gown of white brocade, underskirt of white satin, draped with deep flounce of accordian-pleated chiffon, overskirt of brocade prettily draped with chiffon & orange blossom, white applique on the pouched bodice, sleeves and yoke of Honiton lace, a girdle of chiffon caught with pearls, and a full court train; a Bridal Wreath and embroidered tulle veil were worn, and the Bride carried a shower bouquet of roses, tuber-roses, carnations, & asparagus fern, with white streamers. Miss G. V. McCarthy, Sister of the Bride, the first Bridesmaid, wore a dainty dress of soft cream silk, hat of

cream satin and chiffon, with black velvet and carried a shower bouquet with pink streamers, and wore a gold brooch set with pearls, and a silk Maltese lace handkerchief, the gifts of the Bridegroom.

Fairy McCarthy, the Sister of the Bride, wore a soft pink silk frock, tucked with chiffon insertion, white hat with tucked silk, plumed, lined with tucked chiffon, and chiffon strings caught at the left side, with gold buckle, and carried a Shepherd's Crook, with bunches of pink ribbon and basket of flowers; she also wore Maltese lace handkerchief and buckle, gifts of Bridegroom. Miss E. Catling, another Bridesmaid, wore a pink silk frock, white hat with tucked silk, feathers and chiffon strings, and carried a Shepherd's Crook and basket of flowers, and also a Maltese lace handkerchief, the gift of the Bridegroom. The Misses Molly and Alma Smith (Guildford), Train-Bearers – two little dots – looked charming. They wore cream silk frocks, with short sleeves, trimmed with lace insertion and bebe ribbon, creme silk hats, and carried Maltese lace handkerchiefs. Mr. J. McGregor, of Cue, acted as Best Man. Mrs. Duggan played the 'Wedding March' as the newly-married couple left the Church. The Bride's going away dress was of fawn cloth, strapped with the same coloured silk, and finished with iridescent applique; the blouse was of white tucked crepe de chine, trimmed with lace, and with transparent yoke; black hat trimmed with Paris lace and chiffon. A large number of presents was received, including a gold locket, the gift of the Bride to the Bridegroom, and a Gold Bracelet, the gift of the Bridegroom to the Bride.[27]

Herbert Hoover, as well as being a guest at the wedding, presented them with a beautiful timber dining room table. This table may still be at *Austin Downs* where its history has probably been all but forgotten now.

Social life
J.P. was interested in many charities during his life, and because of Grandfather's love for horses he became a member

---

[27] *Western Mail*, 28 June 1902, page 42

and President of the Cue Racing Club; he most likely enjoyed a little flutter once in a while, I would have thought.

Grandfather was also very interested in outdoor activities and, together with the other station owners around Cue and Meekatharra, they would put on gymkhanas, social days or weekends each year, where the children could compete with each other in horse riding events, play tennis or take part in other activities.

Figure 5: Minnie Meehan   *Figure 5: J.P. Meehan*

The men of course would pit their horsemanship skills against each other, as would the local ladies with their cooking and jam making as well as their sewing skills, from what Mother told us. Everyone was involved in some way and would compete with great gusto. J.P. encouraged his children to become involved as well; grandfather was a great believer in everyone keeping busy.Minnie (J.P.'s wife) was also very active in the local community; as a nurse she was very aware of how important it was to become involved with other women and children in her community.

J.P also participated in the social life around his *Gracemere Park* property at Coomberdale. He was Patron of the football club there for years and got to know their neighbours. He also took an interest in the roads and the provision of telephone lines in the district.

Family Life

Minnie's sisters visited her at *Austin Downs* quite frequently apparently, and her brother Percy at Meekatharra was not too far away.

When they were first married, J.P.'s mother Julia lived with them some of the time, but she died before Grace was born.

J.P. and Minnie raised three children; Arthur, Jack and Grace. My sister Maureen reminded me that our father was one of twins – his sibling died soon after birth I believe. I have little or no information on this child other than it may have been a boy.

Sometime after J.P. and Minnie were married in 1902 our grandfather talked Clementina McCarthy (his mother-in-law) into coming to live with them at *Austin Downs*; that way Minnie was able to take care of her mother. Clementina was a very independent lady although she was blind and could not work.

J.P. made sure that she could move around the house and the gardens by installing a safely rail around the veranda and rails in the hallway and bathroom as well as guide rails around the garden. That way Clementina was able with a little help to enjoy pottering around outside and she was able to help entertain and teach her grandchildren. I am told J.P. showed a great deal of compassion and love towards Clementina during her time at *Austin Downs*.

Clementina had grown up speaking French and was able to teach her grandchildren the language during her time at *Austin*

Downs. Both Father and Aunty Grace could still talk in French as adults.

Sadly, on a visit to one of her married daughters near Northam, our great-grandmother was badly burned when her skirt caught on fire and she died shortly afterwards.

Arthur (our father) remembered both his grandmothers living at *Austin Downs*, although he was only about four years old when he died. Clementina died when he was 17 years old so he got to know her well, especially through the French lessons.

J.P.'s Philosophy

J.P. always talked to his family and friends about his Irish Heritage and that his Irish parents believed very strongly in bringing up their children in the traditional Irish way, tough and with character, which the Irish believed would then stand them in good stead throughout their lives.

He told his family that those days were full of hard work and life was tough, so he taught his sons Arthur and Jack as well as his daughter Grace to take things as they came, to work hard and life would be good to them.

Grace was taught to run *Austin Downs* just like the boys and she was her father's right hand when Arthur and Jack both left home to work and live in America. In addition to *Austin Downs*, there were the *Mt Vernon* and *Gracemere* properties to be considered so there was a lot to keep in mind.

Grace Meehan Harris

Grace Meehan, the only daughter of J.P. and Min Meehan, was born in 1907. From all accounts Grace was the apple of her father's eye and could ride horses and motor bikes with the best of them. She had been named for her mother's sister, Gracie McCarthy.

Known as Gracie to her family and friends, she attended Loreto Convent School in Perth for a while, but her heart was on the station and she returned as soon as she could. However, she also had a busy social round, visiting Perth with her mother and staying with relatives and friends, according to the social columns in the newspapers of the day.

Grace and her parents moved down to *Gracemere Park* each spring for the shearing and she participated in the local social events there. Grace may have met Joan Keamy who was the daughter of a neighbouring property owner and later became her brother Jack's wife.

J.P. and Minnie threw Grace a big 21st birthday party at *Austin Downs*. It was described in the Sunday Times:

> On Saturday evening, October 13th; Mr. and Mrs. J. P. Meehan entertained a large number of friends at a dance at their home, *Austin Downs Station*. The evening was a perfect starlit one and the run across the mulga plain was enjoyed by the visitors.
>
> Mrs. Meehan welcomed her guests in a handsome black georgette frock beaded in red, and was assisted by her daughter, who looked charming in a dainty frock of green georgette and gold lace. The dance room was beautifully decorated with festooned streamers and balloons.
>
> A dainty supper was laid out on the spacious verandah, where the tables was profusely decorated with beautiful flowers. The drawing room was decorated with red and white roses. The hostess and her daughter were ably assisted by Mesdames A.L. B. Lefroy and Price. Mr A. L. B. Lefroy, whilst proposing the toast of the host and hostess, congratulated Miss Meehan on having attained her majority, and wished her many years of health and happiness. The toast was drunk with musical honors. Mr. Meehan responded briefly, after which a return was made to the dance room till 2.30, after which the guests set out for their homes.[28]

When her brothers, Jack and Arthur, were living in America, Grace helped her father run *Austin Downs*. After the death of

---

[28] *Sunday Times*, 28 October 1928, page 11

her mother in 1936, (when she was 29 years of age) Grace ran the household until her brother Jack (who had eventually returned to Australia in 1941) married Joan in 1944.

I was born prematurely in 1942, and Aunty Grace, in addition to her other duties, fostered my sister Maureen who was only a year old then. It was nearly a year before Maureen could come home, and the separation was very hard on both Maureen and Aunty Grace.

I have heard that when Grandfather died, Grace was supposed to inherit a share of the station. However, when it was clear that this was not going to happen, Grace left *Austin Downs* very sad and unhappy. I have no idea if that information is true.

It also has been said that she hitchhiked into Cue from *Austin Downs*; an unheard of action by a lady of those times, or so I have been told. I have no idea if the above is true either but it is a great story and our Aunty Grace would love to read such a story about her. Our aunt would have looked great on a bike.

Grace met and married James Harris (Jim), a fitter and turner by trade. They lived in Geraldton for a long time and then moved to York. Jim Harris died in the 1970s and Grace died in the 1980s. Both are buried in the York Cemetery.[29]

---

[29] Grace and Jim Harris had a daughter, Doreen, who followed in her Grandmother Minnie McCarthy Meehan's footsteps and became a registered nurse, specialising in community nursing (early childhood, I believe). Doreen married Maxwell Crispe, a Perth Barrister, in 1977. Doreen and Maxwell have three children. Ellie is a Veterinarian, specialising in race horses, Kate is a Physiotherapist/Solicitor and works with her father. Simon, their son, is an Architect.

End of an Era

Minnie Meehan, our grandmother, died on the 6th December 1936 from a bowel obstruction. She was a patient at St John of God Hospital in Subiaco at the time.

Grandmother was quite young when she died – only 54 years old – and after a Requiem Mass at St Mary's Cathedral in Perth, she was buried in Northam. This was her childhood home and where her parents were buried.

Mrs. I. B. Meehan.

The funeral of the late Mrs. Ida Blanche Meehan took place in the Roman Catholic portion of the Northam Cemetery on December 8. The late Mrs. Meehan was born in Northam and was a daughter of Mr. and Mrs. D. McCarthy, who will be remembered as old residents of that district. Since her marriage in Northam to Mr. John P. Meehan, of Austin Downs station, Cue, Mrs. Meehan had resided in the Cue district for the past 35 years. Extremely well known and popular, Mrs. Meehan was possessed of a kindly and charitable disposition, which had endeared her to a wide circle of friends, and her untimely passing will be deeply mourned by all who were privileged to know her. She leaves a husband, one daughter and two sons to mourn their great loss. A requiem mass was celebrated in St. Mary's Cathedral, Perth, on Monday and the funeral took place on the following day at Northam, the Rev. Father S. Reidy officiating at the graveside. The chief mourners were Mr. J. P. Meehan (husband), Grace (daughter), Arthur (son), Mr. P. McCarthy (brother), Mrs. G. Morrison and Mrs. N. Hicks (sisters). Mr. W. Morrison (brother-in-law), Mr. O. Hansen (brother-in-law), Mr. W. F. Meehan (nephew), Mrs. Price and Miss M. Dillon (special friends). The pall-bearers were Messrs. C. P. Murray, D. L. Pitt, H. Evans (Nookawarra Pastoral Co.). F. Walsh (Mileura station), W. Gillett (Elder, Smith and Co., Ltd.) and W. Pavy. Among those present were Messrs. J. L. Prevast (Dalgety and Co.), A. C. R. Horn (Union Bank). J. Charles, A. S Price, G. E. Jones, E. Atkinson. E. J. Philline, J. R. Walsh, F. R. Walsh. W. D. Cole, R. W. McKay. P. O'Driscoll, W. J. Morrison, Mrs. P. O'Driscoll and Miss Joyce

Morrison. The funeral arrangements were carried out by
Messrs. Bowra and O'Dea.[30]

My Memories of Grandfather Meehan

Maureen and I would be taken to visit our grandfather, J.P.
Meehan, at *Austin Downs Station* while he was alive. The
station was just out of Cue. We would all go down, once or twice
a year from what I can remember, though I only have a very
vague memory of him, which is sad. My memory is of a big tall
man with a long beard and a very loud voice that frightened the
life out of us when we were little children.

Baby food

Grandfather Meehan also had his own ideas about raising
children. One of his not so good ideas was to start babies – no
matter their age – on meat. When Maureen and then I were
each about three to six months Grandfather told Mother that
she should stop giving us that awful baby food and start us on a
real diet; meaning of course his home-grown meat. As steak was
Grandfather's favourite that is what he thought Mother should
give Maureen and me.

While telling Mother to do this, grandfather (according to our
parents) would be sneaking Maureen and then me small pieces
of steak to chew on. Mother had to come to the rescue of both
Maureen and again myself as she removed the meat and
stopped us from choking. Grandfather could not see what was
wrong and thought that Mother was fussing over us too much.

Watering the babies

Once on a different occasion when we were down from *Mt
Vernon* and staying with Grandfather on *Austin Downs* there
was a heat-wave so our grandfather turned the sprinklers on to
water the garden and to keep everyone cool. This did work a
treat for a little while. Till Mother came out from the house and

---

[30] *The West Australian*, 10 December 1936, page 12

saw what was happening. Unbeknown to our darling grandfather, our mother had decided to put Maureen and I outside, in our bassinets but on the other side of the hedge that grew down one side of the house, where it was well shaded and cool.

By the time Mother came to check on her babies, she found that grandfather had turned the sprinklers on and the water was spraying over the hedge and directly into the bassinets of us children who were asleep at that time. The bassinets were by then filling with water.

From the stories that our father and mother told us over the years, we were starting to turn somewhat blue; Maureen and I apparently thought it was all great fun for a short time. Grandfather's heart was in the right place, but he just thought that Mother and Father should start toughening us both up early. After all, we were Grandfather's first grandchildren.

This happened twice, to my knowledge.

Saying Goodbye to Grandfather

The last time we saw J.P. was when our parents came down from *Mt Vernon* and stayed at *Austin Downs*. While we were there, Father and Mother took us to see our grandfather in hospital.

Maureen would have been about five and I would have been about four I think. Grandfather was ill and in hospital, but he had asked our parents if they would bring Maureen and myself to see him. We were told to be very quiet and to give him a kiss, and to say how much we loved him. That went well and Father was pleased with his children and how we had conducted ourselves, but then things went downhill.

No one was expecting grandfather to speak to us children and if he did it was expected that he would speak in a feeble or gentle way, not in this great big booming bellow of a voice. "Hello girls". From memory, I got out of that room faster than you could say "Jack Robinson". Maureen was not far behind me, I believe.

Grandfather (J.P. Meehan) died on the 21st of December 1946 from heart disease. He was in the Big Bell District Hospital when he died, ten years after the death of his wife Minnie.

He was buried in the Catholic Cemetery in Cue, on the 30th of December 1946. John Patrick Meehan was 88 years old at the time of his death, and had lived in the Cue area in Western Australia for about sixty years.

.oOo.

CHAPTER 3
Arthur Meehan

Arthur Meehan

Our father, James Arthur Meehan (known as Arthur), was born on the 6th September 1903. He was the first child of J.P. Meehan and Minnie (Ida May McCarthy) of *Austin Downs* station. He was named for his father's brother Jim (James Joseph) and his mother's brother, Arthur McCarthy, who was killed in a railway derailment five years before our father was born.

His brother John (Jack) was born in 1905 and sister Grace in 1907.

Figure 6: Arthur Meehan

Our great grandmother was Clementine Langoulant before she married Denis McCarthy, and their back ground was French. The family remember our father being able to speak French, which his grandmother Clementina and mother, Minnie, had taught them. Clementina was blind when she was living with them at *Austin Downs*, but talking to the children in French was something that she could do.

Our father could also speak Spanish which he learnt while living in America. I do not remember him speaking Spanish but Maureen can. I understand from Maureen that our Aunty Grace may have spoken French during her adult life as well.

Arthur was sent to board at St Ildephonsus College at New Norcia. This was probably in 1913 when he was ten, as that is the year it opened. He used to talk about New Norcia and was very happy there I think. A painting of him as a child appears on the inside of their chapel I believe, though I haven't seen it.

After that he went to the Christian Brothers College here in Adelaide Terrace in Perth but never really talked about being there. I think my father might have left school early; maybe when he was fourteen to go back to the station but I cannot remember him talking about it. Father loved being on the station.

Our Uncle Jack (John) Meehan, who was J.P. and Minnie's second son, was named for his father. I don't know where he went to school, but probably the same as my father. Jack left school at 14 to work on the station. However, I think our Uncle Jack may have been a wild boy at heart and he wanted some adventures. When he was 18, he set off to see the world, as recorded in the local newspaper in 1923.

> "Mr. Jack Meehan, second son of Mr. J. P. Meehan, of *Austin Downs*, who since leaving school some four years ago, has been obtaining station experience, has decided to enter the Australian Navy. Mr. Meehan leaves for Perth next week to finalise arrangements which will bind him to a life on the ocean wave for seven years and a further term of five years on the reserve, during which he will be liable to be called up in the event of war."[31]

From what our father told us, Jack soon found it was not to his liking and wanted to get out so he turned to his mother, Minnie, for help. I have heard it said that J.P. – our grandfather – wanted Jack to finish his contract term, but our grandmother

---

[31] *Daily Telegraph and North Murchison and Pilbarra Gazette*, 19 January 1923, page 2

became very persistent that he buy Jack's tenure out. So in the end J.P. paid for Jack to be released from the Australian Navy.

Jack then wanted to follow in his grandfather William Meehan's footsteps, and go to America – much to his mother's horror. However, his father thought it would be good for Jack to travel and experience life and other cultures before he settled down. He also knew that his friend, Herbert Clarke Hoover, was there to keep an eye on Jack and he felt that his son would benefit from the experience.

Jack went off to visit America when he was still under 20 years of age, much to his mother Minnie's horror. Soon after he arrived we are told he joined the American Merchant Navy.

Figure 7: Jack (seated) and Arthur Meehan

Our father found this all rather amusing. He told us that he was very surprised when he heard that his brother Jack had signed up with the American Merchant Navy and even more that he went on to stay with them for quite some time.

I have been told that Jack spent nearly 20 years working in America and it has been said that he applied for and received the right to become an American citizen. This meant that Jack held a dual Australian and American citizenship up until his death in 1962. I am not sure if this is correct as I have not been able to verify this information.

In the meantime, Arthur as eldest son was working with his father on *Austin Downs*, along with his sister Grace. However, J.P. kept the business in the name of 'J.P. Meehan and sons' and apparently promised Grace a share later.

At first apparently Jack was a good correspondent but after he left the Merchant Navy and started moving around on land they lost track of him.

Their mother, Minnie Meehan, eventually became so concerned about Jack that she asked her elder son (our father Arthur Meehan) to go to America to try and find him. At that stage the family had not heard from Jack for quite some time and did not know where he was.

The local newspaper reported on the problem.

> "Since the end of last year no letter has been received from Mr. Jack Meehan, younger son of Mr. J. P. Meehan, *Austin Downs,* by any of his relations. When the last letter received was written, he was then somewhere in Mexico. Usually the wanderer was a good correspondent, and the absence of any message from him for so long a period, has caused grave concern for his welfare. Mr. Arthur Meehan is now on the way to the place at which the last letter was written, hoping to obtain some information as to his brother's whereabouts."[32]

So in August 1929 our father, Arthur Meehan, was on his way to Mexico to search for Jack and hopefully bring him home. Jack was then about 24 years old and had been away for about five years. Arthur was about 26 and they had just purchased *Mt Vernon Station* which he was supposed to be managing. However, our father always told us that it was a great opportunity to travel the cattle areas of the US and Mexico.

---

[32] *Geraldton Guardian and Express*, 10 August 1929, page 2: Personal

This left just Grace with their father on *Austin Downs, Gracemere* and the newly acquired *Mt Vernon Station.*

Uncle Jack was a very handsome man with film star looks. It has been said that during Jack's time in America he met and became friends with Alan Ladd who was an American Screen Actor of some note. It was through this friendship that Jack Meehan said to have been approached to take part as an extra in a number of movies. I have not been able to find any information to verify this. There was a Jack Meehan who appeared in some of Alan Ladd's films but he may be someone else entirely.

Arthur was in America for maybe five years or a bit more, working in the cattle areas of Mexico and the US. From the stories that I can remember him telling us when we were growing up, he worked on ranches in Texas and other states in America as well as in some parts of Mexico while he was there.

There was some talk that either Jack or Arthur had become engaged to an American girl but she died from a ruptured appendix, possibly because of the lack of medical help that was available at that time, and where she also lived, which may have been in a rural area in America.

Arthur returned to Australia in 1936 aged about 33 and with a touch of American in his voice along with a touch of Mexican Spanish. His parents threw a big welcome-home party and invited all the young people of the district. This is where he met Kit Murphy.

Marriage of Arthur Meehan and Kit Murphy

Father and mother met while mother was nursing at the Meekatharra hospital or the Big Bell Hospital at the time. After a rather short courtship they were married. Our mother was Catherine Alicia Murphy and she was known as Kit. She was 27 and he was 33 years old.

Mother's family came from Ireland and they were living in a small country town called Quairading, about two hours away from Perth by train when we were young. It is to the east of York. Mother had been born in Tambelup in 1909 soon after the family arrived in Australia.

The family background was shadowy, as her parents were escaping Irish political turmoils. Kit's father had been an outspoken academic at the university in Dublin and he had been driven into hiding in Australia which made for a difficult life for the whole family.

Her family's story is in Appendix 4, and the story of her Irish heritage in Appendix 5.

Kit had three older sisters to whom she was very close, and a younger brother. Her father died in a work accident when she was 16 and soon after she ran away to Kalgoorlie and started nursing, or so I understand. She lied about her age, and when they found out, they took the money out of her pay until she had refunded all her pay from when she was under age.

Catherine Murphy, Kit's mother, died of a heart attack in 1937, the year after Kit and Arthur were married.

Their wedding was on the 24th of October 1936 at the Catholic Church in Meekatharra. Kitty wore a pink dress and mother's eldest sister, Nell Murphy Partlon was her maid of honour I believe. Before marrying our father and moving to *Mt Vernon* our mother was a registered nurse.

This book is the story of their lives together.

After the Wedding

Arthur's brother Jack was not at their wedding as he was still in America. Then their mother Minnie (Ida May) Meehan died suddenly on the 6th of December 1936, just 6 weeks after that wedding. She was in St John of God Hospital, Subiaco, and Jack was still at that time living in Oklahoma, USA according to the death notice.[33] Minnie had died without seeing him again in all those twenty years.

I can remember Father and Mother telling us that his mother Minnie Meehan had put pressure on J.P. to give *Austin Downs* to Jack as an enticement to get Jack (who was still living in America at that time) to return to Australia. Jack had, to that point, shown no interest in returning to Australia or *Austin Downs*.

Grandfather apparently promised Minnie to leave Austin Downs to Jack if he came back. But to do this, he had to find something to do with Arthur who had already spent a good many years on his father's service. Grandfather offered Father the chance to buy out *Mt Vernon Station* by borrowing the money required from J.A. Father's business dealings were thereafter in the name of 'J.A. & C Meehan', with his new wife as partner.

However, there was a sting in the tail for our father; Grandfather told his son that the loan had to be repaid plus any interest incurred within a certain number of years. I think it was something like fifteen years or thereabouts, which our father paid before we left *Mt Vernon*.

The rest of J.P.s properties were then available to entice Jack back from America. The chance to own the stations outright proved to be the carrot that Jack needed.

Jack returned from America in 1941[34] (when he was 36 years old) to take over the running of *Austin Downs* from his father, J.P. Meehan.

His sister Grace had been assisting her father up to this point, but in the end she left the station to start a new life.

---

[33] *The West Australian*, 7 December 1936, page 1: Family Notices
[34] In the third year of World War II.

Jack and Joan Meehan

In late December 1942 Jack got engaged, and in June of 1944 he married Joan Keamy from *Cardo Station*[35], Watheroo. Joan was the daughter of Les[36] and Ethel Keamy of Watheroo where the Keamy family ran a premier sheep stud.[37]

Grandfather J.P. died at the end of 1946 and Jack inherited *Austin Downs* and *Gracemere* as his mother had wanted.

Joan and Jack lived at *Austin Downs* and had two daughters; Patricia born in 1947 and Pamela born in 1951. They were about the same age as my younger sisters, Judy and Liz and born while we were still at *Mt Vernon*.

Jack decided to simplify the management, sold off the *Gracemere* properties and gave up the stud sheep and cattle. Then in 1962 he died suddenly at *Austin Downs*. He was 57 years old. Patricia was 15 at the time, and Pamela was only 11 years old.

Unfortunately, Jack died without a will and with complex business dealings. Joan and the girls had to leave *Austin Downs*, but with the help of her family was able to start a new life in Dalkeith. Unfortunately, owing to business dealings with

---

[35] *Cardo Station* was not far from *Gracemere Park*. J.P. Meehan and Les Keamy were apparently friends as well as neighbours.

[36] Les Keamy: http://www.carnamah.com.au/bio/leslie-keamy, accessed 16Nov16.

[37] Some of Joan's favourite pastimes were writing poetry and playing the piano. She also loved keeping in touch with family and extended family members, and I was privileged to have been able to talk to Joan regularly while I lived in the eastern states and after I returned to live in Perth in 2005. My daughter Marita was also given the chance to meet her Great Aunty Joan when she visited Perth on her way back from the UK in 1999. Her Aunty – my sister Elizabeth – took Marita to meet Joan. Marita found it a wonderful chance to meet this great Aunty that I had spoken of so often.

Hillside Station, it also left our father in dire straits. I cover that in another chapter.

Sadly, Patricia died in a car accident on the 15th of May 1968 aged just 21 and was buried at the Karrakatta Cemetery.

In 2006 when Marita married Gaven Soutar here in Perth we were lucky to have most of the family at their wedding, but Aunty Joan was not well enough to attend. However, Pam was able to come, and it was a wonderful chance for Marita and Gavin to have all the family together and to meet our extended family as well.

Arthur and Kitty

Meanwhile, Kitty and Arthur made their home on *Mt Vernon* after their wedding. They lived there until 1951 when it had to be sold and we moved to *Hillside Station.*

Father died on the 3rd of December 1973. At the time of our father's death he had just turned 70 and had been suffering from Parkinson's disease for a number of years. He was buried at the Karrakatta Cemetery.

Kit died in Perth on the 11th of January 1977 from heart disease. She was 67 years old and was also buried at the Karrakatta Cemetery.

.o0o.

CHAPTER 4
A Leap of Faith

Our mother Catherine (Kit) Murphy was a registered nurse working at the Meekatharra hospital when she first met the Meehan family from *Austin Downs* and their son (our father) Arthur Meehan.

Figure 8: Kitty Meehan

Mother was known as Kit to her family and friends. After our parents were married they immediately left for *Mt Vernon*, and as I understand from both Father and Mother there was no time for a honeymoon. Father and our mother along with her sister Margaret (Meg) returned to *Mt Vernon Station*, a cattle property that he owned in the Murchison. This was to be the first time mother had been to *Mt Vernon* so it was truly a leap of faith by our mother to leave family and friends for the unknown.

Mother had a great shock when the newlyweds first arrived after their wedding. Mother arrived at *Mt Vernon* expecting the perfect homestead that her husband had told her about; something like a better version of *Austin Downs* homestead perhaps. But it was not quite the dream house that father had wooed mother with stories about. The stories that our father had told our Mother about were quite different. To say mother was somewhat surprised would be true, as well as disappointed.

Mother never recovered from what she encountered on *Mt Vernon*. It was a very basic house. A lovely house from memory but my mother was frightened at living there because there was no one to talk to. No one came to visit and those few who did very rarely stayed over.

Mother must have received quite a shock when she arrived on *Mt Vernon Station* with no electricity; that was to come much later in fact. The only light that the homestead had was provided by Tilley lanterns which used kerosene and there were also kerosene lamps that the family would use at night. There was always a good supply of candles which were a must at night time.

Figure 9: Mt Vernon homestead in the early days

Mother also had to get used to the idea that the main house toilet – which was what was known as a bush toilet – was situated at the far end of the homestead yard, right on the fence line; as far away from the house as possible. Father could not see the problems that our mother saw. My mother claimed that our father was looking through rose-tinted glasses when he looked at the problems that Mother saw. First Mother told Father that he had to fix the toilet. She claimed it was unsafe for children to use, but Father claimed that the children would only fall once and then they would be a lot more careful. Mother was not amused.

To Mother's horror and dismay; worse was come. Father, being Father, thought very little about our mother's reaction to being told that her duties included taking care of the bush toilet. He also informed Mother that she would have to get used to it. Father was not a very thoughtful man when it came to Mother and the fact that she had never seen such an antiquated thing in her life, but she learnt quickly as they both said it was either get used to it or find a bush very quickly.

Now a bush toilet is no more than a hole in the ground, a toilet seat for sitting on and of course behind the door a string with sheets of torn up newspaper. Mother changed that by ordering toilet paper for the station.

It became Mother's responsibility to take care of the toilet's hygiene. This she did by using ash from the wood stoves which she would sprinkle over the waste matter each day; a container of the ash was kept next to the toilet seat in the toilet

Mother also had to organise Father and the station hands to empty the waste matter and remove it as far away as possible from the house every few weeks. Not a very nice job I can tell you. Maureen and I, as we got older, had to help. Yuck! Poor Mother, she must have wondered what she had done by marrying our father who was taking all of this in his stride, of course.

I have suddenly remembered that I was terrified of that toilet because I thought, as a young child, that I could fall in. The seat seemed to be so big and I was small.

Father and mother lived on *Mt Vernon* during the late 1930s to the early 1950s. Her sister, Meg, stayed with Mother for a few weeks to help her settle in and later used to come and stay quite often while they lived on *Mt Vernon*. Meg's husband, Jack Morrissey, would also come and stay.

Mother grew up on a farm near the small town of Quairading and the family lived in a farm house that for that time, in the early 1900s, would have been quite a modern building. Mother and her family had moved into the township of Quairading after Grandfather Murphy died and they continued to live there after Mother married and moved north.

Mother must have looked back on her life as a child, when the towns and cities had a night watchman who came around each night in a specially fitted-out truck to take away the used pan and replace it with a fresh pan for the householder to use. Or those that could afford it put in septic tanks (*Hillside Station* had two septic tanks when we moved there). Then when she had gone nursing, the hospitals that she worked at had septic tanks for the toilets as I understand, so she would have recalled how different it was to what she had to adjust to now.

Because Mother mainly nursed at country hospitals, these hospitals would provide accommodation for the doctors and the nurses. Mother spoke about the hospitals at which she nursed as being quite modern, in terms of the accommodation that was provided for their staff. So even though mother moved a lot between country hospitals during her nursing career, she had never come across the same conditions as she found on *Mt Vernon* as a new bride.

To be fair, the *Mt Vernon* homestead was built in the early 1930s and, for its time, was quite up to date and comfortable; a lot of station homesteads were pretty rough and ready buildings as money was always in short supply so what was available went into running the station.

It did not take long for Mother to realise that she had a major and rough road ahead.

She first worked out that she would have to start to make plans from top to the bottom of the homestead; plan after plan was tossed aside before Mother was happy. Meanwhile, Mother was in her seven heavens. She agreed that *Mt Vernon* homestead was a lovely homestead indeed, and with some changes it would be perfect.

She declared "Just let me at it."

This was when the newlyweds first returned to *Mt Vernon*. Father was frozen in his boots at the very thought about what changes his brand new wife might want to do to his beloved homestead. The station that he had made and treasured with his cattle. What would Kitty do? She made plans, thought it all through, and then got at it.

When mother first moved to *Mt Vernon* she had to learn how to make her own bread, as well as learning to make the yeast for the bread using potatoes. Using potatoes to make the yeast could be quite troublesome as the lid would often get blown off. Then mother would have to start all over again.

> The potato yeast receipt; one large cooked potato mashed, add one teaspoon salt and one teaspoon sugar, mix in four cups potato water and mix well, place potato mix in a large glass jar. Do not put the lid on the jar, and place in a warm place in the kitchen, leave for a few days for mix to foment or until mixture starts to develop bubbles.

Mother was unable to purchase yeast from Elder Smith in Meekatharra during those early years on *Mt Vernon*. So mother felt like she had achieved something when she was finally able to purchase ready-made yeast to make her bread. Mother called it proper yeast.

She also had a pressure cooker which was just as cantankerous and would often blow the lid off as well, much to Mother's annoyance.

Fanny and mother also used to make soap from fat from the slaughtered animals and mix in caustic soda; given time this would turn into a type of washing soap and was used for washing difficult articles.

The cattle musters would be carried out about three times a year and would last for weeks – anything from six to eight weeks at a time – and after the muster the cattle were then taken to Onslow or Meekatharra to be put on a ship or train to go to the markets in Perth.

Figure 10: Kitty (left) and Aunt Meg at mustering camp, Mt Vernon Station

Mother and her sister Meg went out to the mustering camp a couple of times; that was before mother had us children and when father had the ute. I think they ended up being the cooks while they were there. That was another story because both Mother and my aunt had to learn to cook with a camp oven which was a large cast-iron pot and they cooked mainly damper and stew type things from what my mother used to tell us.

It was during one of my Aunty Meg's visits that mother and her sister tried to cut off father's moustache while he took a nap. Our father was not a happy chappie, as he had been growing his moustache for many months and he was so proud of it but now it was a mess because mother and her sister had only managed to cut off half the moustache. Anyway father refused to allow them to cut it all off so father went around with just a half moustache till the cut off half grew back, so our aunt and Mother told us.

Babies at *Mt Vernon*

Mother had lost her first baby (Mary Philomena) soon after birth in 1938, so she was four years on Mt Vernon before she had children for company at home.

In many ways I believe that Maureen and I must have been quite a challenge for our mother. Maureen was born in July 1941; I was born eleven months later. From what our mother told me I was a premature baby, born in June 1942.

While Mother was in Geraldton awaiting my birth she started to show signs of going into early labour, so our father's sister Grace came down to Geraldton from Cue and took Maureen back with her to *Austin Downs* station. Maureen would have been less than 11 months old at the time.

Mother always said that I was a difficult baby; and so she felt that she never really connected with me. I wonder if the same story might also apply to Maureen's birth as well, because it made a lot of sense when I looked back at our relationship with our mother.

From my memory Mother was quite different when she had Judy and Elizabeth, particularly Judy. She seemed more comfortable and more connected to their care. Judy was her blue eye blonde baby and Elizabeth was her baby for some time.

It was some time before Mother returned to *Mt Vernon* with me; she may have gone down to be with her sister Bridie, in Quairading, where she must have stayed till she felt more able to cope with a new-born.

During this time Maureen stayed with Grace and grandfather (J.P. Meehan) at *Austin Downs*. Maureen must have been at *Austin Downs* for a number of months; possibly six months or more. It must have been in January 1943 that we all returned to the station. I understand that Mother's sister Meg came back with Mother to help look after us on the train and also to help Mother. I think Grace (Father's sister) came to *Mt Vernon* at some stage as well, but that was not a happy marriage between the sisters in-law. However, Grace ended up staying for some time, from what Father and Mother said.

I suspect my early arrival also affected mother's ability to reconnect with Maureen after their enforced and long separation. Maureen had bonded with her Aunty Grace while living with her at *Austin Downs*.

To make things more difficult for everyone, Father would be gone for weeks at a time, mustering or droving cattle to the markets. I imagine this just made things a lot worse for Mother.

Isolation

Father had painted a magic life to my mother of sleeping out under the stars. Of magic nights watching the roaring fire, of picnics by the river. It was a dream that my father loved but for Mother this just spelt being alone, of being left alone at the homestead for weeks at a time as my father and his men did the mustering. Mother just dreamed of being with her sisters and her family and her nurse friends and other friends.

So my mother started to beg and pester, saying to my father "I will go mad with being alone". Mother told my father that she would go mad without seeing people. Mother never felt safe. My father never had the same fears as my mother, and never really understood what made my mother so frightened.

It was a very hard and lonely life for the women folk who chose to marry men from the North West as they had to learn to adjust to a very different way of life to what many of them had been used to. Being so far away from other people meant that these women became very isolated. Even a visit to a country town for a break and a chance to meet other women from time to time was a big event as it could take three to four days drive to get there and that was over unmade roads. More like dirt tracks actually.

Father apparently tried to teach Mother to ride a horse when they were first married. Mother did try to learn to ride but my father yelled so much Mother nearly fell off the horse. So Mother and Father both yelled to say "No more horse rides".

He even attempted to teach her to drive a car as well, but his yelling was too much for her. In fact during one lesson, his yelling was so bad that she ended up running the car into a tree. In the end, mother refused to get on a horse or to drive his car while he was around, or so she would tell me later in life.

Figure 11: Arthur Meehan with his first car

Our closest neighbours were a full day's drive from *Mt Vernon*, so father always packed swags and food for an overnight trip just in case things did not go according to plan. It was not unusual to have the car break down on one of these long trips, so Father and Mother had to be prepared for delays along the way.

Mail was delivered about once a month during the 1930s and up to 1949; this was when an airstrip was built about half a mile from the *Mt Vernon* homestead. At the same time a Wind Generator was installed to give the station electricity for the first time. With electricity, the station was able to upgrade from the Pedal Radio to a more modern and easier to operate model. The new radio ran on batteries which had to be charged every day; this was the reason behind the Wind Generator being built and placed just outside the radio room. Mother was very happy when told that she could toss the old fashioned pedal set which until then was the only means the station had to communicate with the outside world.

After her initial shock; mother went on to become a very self-reliant lady, and with the help of her Aboriginal friend Fanny, made a lovely home for her own family and for the Aboriginal families that lived on *Mt Vernon* and worked for father.

Our family

Arthur and Kit Meehan had six of us children.

Mary Philomena born 25th July 1938 at St John of God in Subiaco. Mary lived for a short while and passed away on the 25th of July 1938; Mary Philomena is buried at the Karrakatta Cemetery.

Maureen Catherine Meehan Stubbs was born on the 3rd of July 1941, at St John of God Hospital in Geraldton. Maureen now lives on a farm in Bindoon W.A..

Margaret Frances Virginia Meehan O'Shannassy (known as Virginia) was born prematurely on the 10th of June 1942 at St

John of God Hospital in Geraldton and now lives in Rockingham W.A.

Judith Anne Meehan de Maniel, born on the 14th of August 1947. Judith was born at St John of God Hospital in Geraldton. She died at home in Sydney on the 18th December 2009 and her ashes were returned to *Hillside Station* in the Pilbara on the 28th of September 2010.

Elizabeth Carmen Miranda Meehan Krieger was born on the 7th of March 1949 at St John of God Hospital in Subiaco and lives in Willagee W.A.

John Patrick Meehan was born on the 11th of May 1954, prematurely, while Mother was back home on *Hillside Station* and lived for a short while before passing away. John Patrick Meehan is buried on *Hillside Station* near the rock grotto and sun dial; Mother and Father wanted John Patrick to be able to see the passing of time even in death.

.oOo.

CHAPTER 5
Mt Vernon Station

Developing Mt Vernon Station

Our father, James Arthur Meehan (known as Arthur), took up a Pastoral Lease in the Murchison area in 1929.

> Messrs. J. P. Meehan & Sons, who recently acquired *Mt. Vernon* Station, are removing all their cattle from *Austin Downs*, and sending them to the former station, which is to be 'fully' stocked with cattle. *Austin Downs* will in future be solely a sheep station. Mr. Arthur Meehan will take charge of *Mt. Vernon*. [38]

Before Father took out his lease on the land, he surveyed and pegged out over one million acres which he believed to be prime cattle country. A very important feature that our father would have been looking for was water, so during his search he concentrated on finding what water was available. This included searching for ground water (an Artesian Basin) so that wells could be bored. As well, he looked for gorges that had billabongs and he also checked to see what rivers ran through the property and if they were available to him. After establishing that there was a good water source father decided that *Mt Vernon* would be ideal for a pastoral property. Father and Grandfather became very interested in setting up a cattle station after studying the results of father's findings, but it could be difficult country.

Serious Effects of Floods – Stock in Good Condition

> Mr. J. P. Meehan, principal of the firm of J. P. Meehan & Sons, returned last week from a trip to *Mt. Vernon*, where his son (Mr. Arthur Meehan) is in charge of the *Mt. Vernon* cattle station. The trip had been attempted some time ago, but after reaching *Bilgun Station* the heavy rains prevented

---

[38] *Geraldton Guardian and Express*, 9 August 1929, page 1: Cattle for Mt Vernon.

further progress. After remaining at that station for some weeks, Mr. Meehan returned to *Austin Downs* to await a more favourable opportunity. On this occasion, there was no rain, and the roads were dry, hard and very rough; so rough in the river beds and tributaries of the Murchison, Gascoyne and Ashburton Rivers that travelling was far from easy. "The floods must have been terrific," Mr. Meehan states, "as for miles on each side of the rivers mentioned the earth had been swept away, leaving deep ruts where roadways existed, and the river beds had been so altered that the original crossings could not be followed. The flats near the rivers and creeks, denuded of earth matter, were very backward in growth, but in the hilly country there was abundance of feed, and stock was now fit to travel to market. Mobs of bullocks and flocks of sheep are already on the road to the nearest railway point, and as the season is a good one over a large area, more movements are expected, amongst which a consignment from *Mt. Vernon* will shortly be on the way. As surface water does not last for any time, and it is reported the stock route wells are not in good order, late consignments are likely to meet with trouble.[39]

J.P. Meehan, our grandfather, was very involved with what Father was planning as he had provided the finance to our father after feeling rather guilty that his wife had pressured him into leaving *Austin Downs* to Jack, the younger son.

I have heard it said that father is still highly regarded in the area for his knowledge and skills both in horsemanship and cattle breeding. During our time on *Mt Vernon* father purchased two more leases which the new owners went on to sell sometime later after we were gone.

After taking out a Pastoral Lease on the land that he had chosen, Father (James Arthur Meehan) named the property *Mt Vernon*. His plan was to establish it as a prime cattle breeding station. Father lived and worked in America for about five or so years I believe. While there he worked on cattle ranches so that he could learn the cattle business inside out. Father's long term

---

[39] *Geraldton Guardian and Express*, 2 June 1934, page 5

plan was that when he finally returned to Australia he would have his own station where he would breed quality shorthorn hereford cattle, and this he did during his time on *Mt Vernon Station*.

Mt Vernon Homestead

Mt Vernon was a beautiful homestead when we lived there. I am not sure who designed it, so below I have tried to remember what I can and tell you what the house was like.

Arthur and J.P. Meehan employed two Italian stonemasons that Grandfather knew to build the homestead. After it was completed, the stonemasons then moved to Perth where they found employment helping to build St Mary's Catholic Cathedral.

Figure 12: Mt Vernon homestead in later years

The *Mt Vernon* homestead was built from stone found on the property and had a corrugated tin roof and concrete floors. Father had gone to a lot of trouble when it came time for him to build his house. Time was taken to select a large clay pan area suitable to build his house and put his work sheds which were set back from the house. This group of building also had quarters for the workmen as well as an ablutions block.

Father also had a fence built around the perimeter of the house to keep us children out of trouble. Outside the house fence father had put in a hitch rail where he and the men could tie the horses up while they were being saddled and the pack horses were made ready for the cattle muster. This meant hooking the pack horses up to a large wagon or dray as this was the only means available at that time. Later father bought a ute – and I think he bought a truck as well – for moving food or equipment that was going to be used during the long weeks ahead.

The homestead was designed to take a large family and to be comfortable; cool in summer and warm in winter. There were four very large bedrooms which ran off a long central hallway that separated the bedrooms from the dining room and the lounge areas. The dining room was a large room with a connecting door into the lounge room. There was another doorway from there into the hallway. The lounge had a large open fire place which the family used in winter. One of Mother's most treasured memories was that after we children were settled in bed, Mother and Father would sit in the lounge room and Father would read to her – either poetry or from the classics – they both had a great love of reading. Then they would retire to their bedroom which was next door to the dining room.

The ceilings in the main house were pressed metal but in the kitchen and office area Father had used corrugated iron. The floors were either stone or concrete, because of the white ants.

A bathroom was built on to the side of the veranda. That way the water from the bathroom could be channelled out on to the garden which was just outside. Maureen and I loved to watch our father when he went to have a shave because he used the old fashioned cut throat razor blade which he had to lather up on a leather strop or belt before he could shave. We thought that was very neat as our father would let us soap up the belt for him and then he would pretend to shave our faces. Mother was not that impressed as she thought – and rightly so – that the two of us would try and shave one another as we were known to get into all sorts of trouble.

There was a second bathroom and this was situated at the toilet end of the homestead next to the garden and chook yard. Father or Mother had planted this area with bamboo and when it grew up it became a thick and lush bamboo grove between the garden and chook run.

Because of the growth of the bamboo it did not take long for us children to claim this as our very own hide-away. What an ideal place for children to play, among the thick and lush foliage of the bamboo. We loved to hide there and play our make-believe games for hours on end. Unfortunately for us children, our parents decided for some reason to remove the bamboo grove. So it was not long before Father and his men had chopped out the bamboo and burned it and the area where the bamboo had grown as well, just to make sure there were no unwelcome visitors lurking. As little children we were heart-broken to find we no longer had a special place to live out our make-believe adventures.

The chook yard and the garden were set well away from the house and Father and the staff used this bathroom mainly, but I loved to snick in and have a shower because the water was always warm and the shower head was quite large. But if I was caught by Mother or Father I was in trouble.

I can remember, one year, one of the cows had a calf which she rejected and mother put the calf down in the veggie garden just behind the bathroom. Mother tried to take care of the little calf as Father was away. She did the best she could, but one morning Mother found that the calf had died, and she was left with getting rid of the dead calf. Mother had Maureen, Fanny and myself helping her pulling the dead calf away from the house till Mother thought it was far enough away; then we covered it with stones.

The house was surrounded by a very wide veranda. This was used by the family during the summer months, when we would all take our beds outside to sleep as it was too hot to sleep in the bedrooms. Sometimes we would all sleep out on the lawn which was fun.

The front of the homestead was always referred to as the 'main entrance'. Here there was another group of buildings and a large and spacious kitchen with a stone floor, two wood-fuelled stoves and two large tables as well as various cupboards. There was a sink for washing dishes or possibly two sinks. I am not sure.

This was where the family and the workers would have their breakfast which usually would be around five o'clock in the morning. Breakfast was either steak and eggs or just eggs and loads of toast; our father loved to add chillies to the steak when he was cooking breakfast for the men.

Lunch and dinner were eaten in the dining room, or the kitchen if Father had a lot of staff working on the station. The kitchen was where Mother did all the cooking, winter and summer.

At some stage Mother and Father purchased two kerosene-run refrigerators. I am not sure what year this was and she had one in the in the kitchen and one in the dining room.

Outside the kitchen was the laundry. This was open to the elements and had two concrete troughs which held a scrubbing board and a mangler (wringer). There was also a copper for boiling linen and the whites (as mother referred to the underwear and tea towels). The copper was heated by wood chips.

Figure 13: Kitty at the main entrance of Mt Vernon homestead

Mother and Fanny would do all the washing by hand back then, as we had no washing machine. The mangler (wringer) was used to remove the water from the washed article and this was done by hand. To use the scrubbing board, you had to rub the clothes back and forth and up and down, or boil them in the copper. Maureen and I were given the task of stomping up and down on any blankets that were in the trough.

This was also where Mother and Fanny made soap. The soap was made from rendered down fat and caustic soda and used for washing clothes and bed linen.

Next to the kitchen was a large room that Father and Mother used as an office and storeroom. Till 1949, the only radio communications that the family had with the outside world was a pedal radio that you had to pedal like you would when riding a bike. That was the only way you could raise a signal on the shortwave radio. This was quite a chore and Maureen and I were often given the job of pedalling – and pedal we did – I think mother used that as a means of keeping the two of us out of mischief.

While we were pumping the pedals, mother would talk to the radio operator in Meekatharra. The radio was used for sending messages, ordering goods for the station and keeping in touch with the Royal Flying Doctors service.

Father used a section of this large room as his office, and it was also where Mother set up her emergency nursing station. This consisted of a locked medical chest which was provided by the RFDS (RFDS is short for Royal Flying Doctors Service which I will use from now on). It was the responsibility of the station owner or manager to keep the medical chest locked and up to date and to make sure that everything was there that Mother or the station people would need.

As well, in this area was a storage room where the supplies that the station would need were stored. We had supplies delivered once every six months and that could be the camel train or in some cases a bullock wagon owned by a Mr Farbor from Marble Bar. I think that was his name.

This consisted of a large dray being pulled by up to six bullocks. This would later change to trucks delivering supplies about every six months or so. Mother or Father would have to order what they needed over the radio. Mother would also buy clothes and Christmas and birthday presents for us children; some Christmases the presents did not arrive on time and Mother had to try and explain to us children that Santa was running late that year. Mother would have Ahern's and Boans as well as

Foys send her catalogues during the year so that she could send in her order on time.

Father had also build a large cool room for our perishables. This was built from spinifex and corrugated iron. The spinifex was stuffed between chicken wire and tin was used on the roof. To keep the room cool, water was dripped down the sides into the spinifex walls. Mother also made good use of a Coolgardie safe for keeping meat in.

The other side of the kitchen there were bedrooms for visitors.

Some distance from the homestead were the station hands' quarters (bunk house), as well as a harness and saddle room and a storage room for the stock feed. Next to that room, Father had a special room where he did repairs on the saddles and harness.

There was also a blacksmith's shed where work was carried out on drays and buggies. Later on this included any motor vehicles that the station had. Set away from these buildings was a locked room where father kept the station's poisons. There was also an area where Father stored the homestead's fuel.

Not far from the men's accommodation, Father had put in a pig run as he had great plans to raise pigs for their meat; it became one of my jobs to feed the pigs each day.

There were also a number of other rooms and sheds. All of these were known as out-buildings and were constructed from local timber (trees) and corrugated iron (tin). Outside one of these sheds, Father had left a broken down station vehicle, possibly a Ford, that Maureen and I and the Aboriginal children were allowed to play on. We would spend hours climbing all over this old car.

On one side of the homestead was a large flat claypan area that father used when moving the mustered cattle down to the homestead stockyards. They were set some distance from the homestead.

These stockyards were where father did the branding and ear marking, as well as the slaughtering of beef which was needed for the household. Needless to say I always got the job of helping Father in this task. Maureen always managed to find somewhere to disappear to. In retrospect, it is a miracle that I never became a vegetarian after all that.

Father also had an area set aside for milking the house cows; this was a job that we both liked, as it meant that we had our Father's full attention for a while at least. Maureen also managed to find a way to amuse herself as well; this meant that I would find myself being sprayed with milk from one of the cows when our father was not looking. One thing that our father allowed us to do while under his care was to ride the cows around the yard while they were waiting to be milked; they were so quiet that we could climb on and off as much as we liked.

While our parents were still living on *Mt Vernon* they purchased a house in Daglish (a suburb of Perth) at 29 Stubbs Terrace. This is where we would stay when the family went to Perth for holidays or for business. The house was sold in 1952 or 1953 I understand.

Fanny

Fanny – our mother's friend and our nanny – was a very important part of our lives, from when we were just little babies to when the family left *Mt Vernon* in 1952. Fanny acted as a confidante to mother as well as helping with us children; she was a very special lady to us all and we missed her dreadfully after moving to *Hillside Station*.

There are no words to use that would tell the reader just what a really special lady Fanny was. To tell you just how important to us children and to Mother, Fanny was the rock. She helped in everything and she made us feel safe. Importantly, she was Mother's friend and her trusty confidante; the only person that Mother went to when Mother lost her first little daughter Mary.

She was very fragile for many months after Mary's loss and again it was Fanny that gave comfort to Mother and our father during this sad and painful times. Fanny spent many afternoons just holding Mother and letting her cry. Fanny also made Mother laugh as well.

Mother was apparently over the moon when she found that she was having another baby but she was frightened that maybe she might lose her baby again like the first time. Mother found that time very stressful so yet again Fanny stepped in to help.

However, Mother did not take to having a baby, so Fanny kept Maureen and I out of trouble when she could.

Mother and Fanny also did fun things together and with us kids tagging along. Things like going for walks and gathering native plants to use when Fanny was going to cook. I think Fanny most likely taught Mother to cook. Mother always said that she found it hard cooking for our father and us children. However, she said cooking outdoors was a hoot because Fanny would go along on these camp-overs and if we were to believe Mother, they could be fun if the ladies could cook as well.

Now I can tell you that Maureen and I were the most perfect children but the truth was that Maureen and I loved getting into mischief if we could. Even that would be stressing the point in that we were into everything that we could get our hands into. We were a handful and Fanny was the only one to keep us under control. Most times! We would do anything if we got half the chance. Like picking Mother's carefully looked-after tomatoes, to Maureen and Jerry rounding up the chooks. Maureen and I would try anything. We nearly put me into a lot of trouble after Maureen had put a dog collar around my neck and then pulled it tight. It was Fanny that saved me. From what our mother and father told us, I was very close to not making it. But that was what Maureen and I were like; full of beans and we would try anything once if we had half the chance.

Fanny and Mother also joined forces to help raise Tadgee after she was found alone under a bush when she was a baby, not long after I was born.

When we moved to *Hillside Station* it was with the understanding that Fanny was coming with us. When we did not find Fanny at *Hillside Station* Maureen, Judy and myself were devastated. Worse was to come. Mother finally found the strength to tell us that Fanny had died at the Onslow Hospital. We were all in shock, we begged father to go and bring Fanny home to *Hillside Station* so that we could look after her. Mother said that she passed away about 1951 or 1952 in the Onslow Hospital from possibly stomach cancer. I can remember being torn and cried for days and days.

The news got worse when we found that Jerry and Tadgee had not come either. We were told that the family of Tadgee and Jerry took them and looked after them but we never heard or saw them after we moved to *Hillside Station*. I was torn apart when I realised that when we left *Mt Vernon* it would be the last time we would be together.

Fanny was a very special lady that only comes into a family just once. As children, Maureen my younger sister Judy and I, we were truly blessed to have had such a lady to love us so much, and for us to have been able to love her back as well.[40]

Improvements at Mt Vernon

Electricity came to *Mt Vernon* in 1949 when an airstrip was built for the RFDS to use; thirteen years after Mother moved in and only three years before we moved on.

This meant that we were introduced to the wonders of a plane service once a week. The mail plane would bring our mail and any perishables that Mother ordered as well as what Father

---

[40] Liz was born just before we left Mt Vernon and so unfortunately she didn't get to remember Fanny.

needed for the station. It also meant that we were not as isolated from the world and could travel to Geraldton or Perth and even Meekatharra, if and when the need arose.

I think it was the Main Roads people that built the air strip and they also built Maureen and I a tennis court. This was interesting as there were no fences to keep the balls from going bush; needless to say Maureen and I lost a lot of balls, but we had a lot of fun as well. Ada (our governess at the time) was given the task of teaching us how to play tennis.

The RFDS made the decision to up-grade the Pedal Radio at the same time as the air strip was built. The new radio was more modern and easier to use which was a blessing in emergencies, but it was run off batteries. To keep the batteries charged, Father had to have a wind generator installed which meant that the homestead now finally had the wonders of electricity to brighten our nights.

Leaving Mt Vernon

Unfortunately, father developed serious health problems from the minerals in the artesian bore water which was used for our domestic water supply on *Mt Vernon*. Its mineral content was too high for our father's body to tolerate and he became quite ill, so finally Father and Mother realized that they would have to sell *Mt Vernon* and move somewhere else.

Father sold *Mt Vernon* to the Panazzi family.

"Pastoral Property Sold For About £18,000

Mt. Vernon cattle station, about 200 miles north-west of Meekatharra, changed hands this week at a price believed to be in the vicinity of £18,000. The property has an area of about 886,000 acres, and the Ashburton River runs through

73

it. Formerly owned by J.A. and C. Meehan, who first
acquired the property under a Crown lease about 1929..."[41]

Captain, an Aboriginal stockman[42] and a very good friend of
Father's, stayed on *Mt Vernon* as head stockman after we left
and I understand he was the manager there for quite some time
as well. Sometime later Captain bought his own cattle station I
believe.

We left *Mt Vernon* in mid-August of 1951. Maureen was 10, I
was 9, and Judy and Elizabeth were 4 and 2 respectively.

"Mr. and Mrs. A Meehan who have recently disposed of their
property "*Mt Vernon*" were passengers for Perth on
Thursday's plane. Mr. Meehan will be returning shortly."[43]

Father then bought a sheep station in the Pilbara – *Hillside
Station* – from Doctor Gillespie. I believe that was his name[44].
Our family moved to *Hillside Station* in 1952. The Rieck family
then owned *Mt Vernon* for about forty years I understand.
Shane Rieck and his family sold *Mt Vernon* after a terrible
aircraft accident on the property. I understand *Mt Vernon* was
sold again in 2014 or 2015.

.oOo.

---

[41] *The West Australian*, 4 Aug 1951, page 4:
http://trove.nla.gov.au/newspaper/article/48983927

[42] Unfortunately, I do not remember Captain's full name.

[43] *Northern Times* (Carnarvon), 16 August 1951, page 3: Onslow News Items

[44] Dr Leslie Thomson Gillespie

CHAPTER 6
Hillside Station

Hillside Station was first established in 1880[45] by John and Emma Withnell, who sailed north, in 1864 on the *"Sea Ripple"* and landed at Tien Tsin with their household items and cattle.

They spent their time while at Roebourne looking around the district and it was not long before they were again on the move looking at land till finally they found what they wanted and settled on *Mount Welcome* and *Hillside Station* (then known as *Shaw River Station*).

Emma Withnell is considered to be the first European woman to settle in the North West.[46]

Figure 14: Pilbara landscape and road (Photo Bree Krieger)

---

[45] Heritage Council: Hillside Station Homestead – http://inherit.stateheritage.wa.gov.au/Public/Inventory/Details/5b2182c3-bd66-4f2a-89e8-595221455ea9

[46] Taken from APAC "The Arrival of Europeans in the Pilbara" page 20

Here I have described the homestead and outer building on *Hillside Station*. I am relying on my memory here so I may not have got things quite right. Please bear with me.

The Homestead at *Hillside Station*

Hillside Station's homestead was very well designed. It was set on about three to four acres, if my memory is correct. If you stand at the front gate and look towards the house, you will see the Shaw River which is on your left, the house is in the middle between the river and the ranges running past the homestead; these ranges have the most beautiful colours which keep changing throughout the day.

Figure 15: Hillside Station homestead

Looking down from the front gate you will see on the left, at the far corner of the homestead grounds, the stockyards which were always very busy during my family's time on *Hillside*.

The homestead grounds consisted of the homestead (house) itself as well as work stations or work sheds. These were large rooms, built using galvanised iron for the roof and pressed or rammed concrete walls.

The house was build using pressed or rammed concrete for the double walls. There was a cavity separating the two layers of concrete in the middle. The roof was made from galvanised iron and the ceilings – from memory – were made from pressed metal. From memory all the floors in the house and the outer buildings were concrete because of the white ants problem (termites). There were no floor coverings, just inside mats.

To describe to the reader the buildings that supported the running of the station, I will stand at the front gate and look down towards the house. You will see on your left a group of work rooms. The first room was used as a room to store the stock feed, next to that was where our father kept poisons and dingo traps – this room was always locked.

Moving along the building we come to a fuel room where kerosene and petrol were kept for use by the station owners. Next to the fuel room there was parking space for vehicles. The next two rooms were stockmens' bunkrooms (bedrooms). After that there came a harness room. There was also parking for the station's truck and a caravan.

Down the middle of the yard and to your right our parents had made two rather large garden beds and planted gum trees and impatiens flowers. On the other side of the yard next to the gardens there were more work sheds. Just in from the front gate on the right was a large blacksmith's work room – this was where our father did all his own blacksmithing.

From memory, I think this room and possibility the generator room were the only work rooms that had dirt floors. This was done for safety reasons and to reduce the risk of a fire when he was working in the blacksmith work area.

There was an extra bedroom here for stockmen to use. We also had a large storage room where the station's supplies were kept. There was another room next to the storage room but I cannot remember what it was used for – possibly a saddle repair room – then around the corner from those rooms was a large room where the two generators for the homestead's electricity were kept. Next to that was a shower and toilet for the staff.

My sister Maureen and I were put in charge of running the generators if our father was away or was late getting back from working away from the homestead, and it was a very difficult job for two young girls I can tell you as we had to refill the petrol tanks of the generator while it was going during the evening. The generator was turned off at about nine o'clock at night. Across the open space from the generator room was a large kitchen and meals area for both the family and the workers. There was also an outside area next to the kitchen where we could sit and have meals during summer evenings.

The kitchen was large and well planned. We had a large wood stove for all of our cooking needs, be it winter or summer. The stove was not far from the main entry to the kitchen.

In the centre of the kitchen there was a large timber and metal-sheeted table for preparing and eating meals. It is hard to believe now, but back then you could fry an egg on the table's metal top if you were so minded, as long as you did it in the middle of the day during summer. Our father was said to have tried doing this for a joke and I think to tease our mother, who was quite impressed. One wall was covered in floor to ceiling shelving. Mother also had two large cupboards to store breakables. Along one wall was a bench where bags of flour and sugar were kept during use.

A large window opened out over the washing-up area. Next to the kitchen was a bedroom for the cook who, during busy times on the station such as shearing, would come out to *Hillside*

Station to help out. The area set aside for washing dishes or when needing extra space was very generous, from memory.

There were three doors leading to and from the kitchen, plus the two doors from the cook's bedroom with one of those doors opening to the yard outside.

One of the doors from the kitchen gave us access to the outside laundry. There was also an area just outside the kitchen which was used to store the chopped wood. The wood was used as fuel for both the kitchen stove and the wood-fired copper. The copper was made from copper metal I believe.

The wood was stored in a forty-four gallon drum which our father had cut in half. Our father always kept the drum full of wood for our mother to use in the kitchen and also the laundry.

Back then bed linen, underwear and tea towels were always boiled in the copper before being hand washed. Our mother bought a petrol-run washing machine a few years before we left *Hillside*. While we had the washing machine for washing, we still boiled the copper up for hot water. Father was quite funny about who could use the petrol-run washing machine and would not allow our mother or us children to use the machine if he was not at home. Not sure why, looking back.

In front of the laundry and just outside the kitchen and along the path leading to the house; mother had planted bougainvillea bushes. They gave colour and were tough and indestructible. Mother had also planted bougainvillea bushes and the impatiens flowers in front of the house. They added colour and gave some protection from the late afternoon sun. There was also a Jacaranda tree and a Poinciana tree on either side of the path leading to the house, and then at the other end of that garden bed were two very large gum trees. I think from memory they were flowering gums which were beautiful when in flower.

Figure 16: Ranges near Hillside Station (Photo Bree Krieger)

The placement of the homestead was interesting; the original owners had built the homestead with the Shaw River on one side and the beautiful ranges to the other side.

Former owners had also planted oleander trees down both sides of the house. At the other end of the house there were also large flowering gum trees, as well as the odd oleander which seemed to survive all weather conditions.

The homestead had six bedrooms, two bathrooms, a lounge room and an office. Set away from the main house were the kitchen, laundry and another bedroom for the cook. There was a very large dining room in the main part of the house, which comfortably took the two large dining room tables that our mother brought with her from *Mt Vernon*.

Figure 17: Shaw River from Hillside Station homestead

There was also a cool room, made from spinifex with water running down the sides as well as water sprayed on the roof. The walls and ceiling were made from fine chicken wire and then stuffed with spinifex. This room was used mainly in the summer, and was set away from the main house.

Away from the house but near the office were two more spinifex rooms; one was for home butchering, and the other room was for storing potatoes and onions which were bought in bulk.

There was a very wide veranda which ran right around the house. During the summer we would take our beds out onto the veranda at night or even onto the lawn where it was a lot cooler and we could sleep. Baby bats could sometimes be a problem as they liked to fly around at night and sometimes land on my pillow. They were harmless but could give us a fright when they landed.

There was a large water tank, next to the house office as well as another tank which was on a kind of a stilt arrangement. This tank was mounted above one of the bathrooms. Water in this bathroom was always warm during winter and summer and the shower in this bathroom was great for using during the hot summer days.

Figure 18: Hillside Station homestead verandah

Stockyards

The stockyards were built close to the main homestead on both stations and as well there were a number of other stockyards spread around both properties. All the stockyards on the two stations were constructed from materials easily found on that station. The yards were all built in the traditional manner, as far as I know. Here I have tried to describe what that was.

Posts and rails were made from the trees that Father (and the men he had working for him) had chopped down and shaped. The gates from memory were made from both metal and timber and all this work was done by hand. Fences were also made using trees chopped down and star posts made from iron I think.

The stockyards were always built near a main paddock; either near the homestead or if they were away from the homestead then the yards would be built near a water source. The homestead stockyards had two main yards; one for herding the cattle and the other was used to separate the animals being worked on and those that were waiting to be done. These stockyards were used for both the cattle and for breaking in father's stock horses. All stockyards were built the same, strong and sturdy.

Gardens

Next to the stockyards was a very large chook run. Needless to say, over time, the chooks were eaten, especially for Christmas dinner. Our mother had turned the land next to the chook yard into a large vegetable garden where she would spend hours each day tending to her vegetables. I can remember her growing tomatoes, carrots, peas, beans and she even managed to grow water melon and rock melon. I think that is where our sister Maureen gets her green thumb from.

Figure 19: Shaw River in flood

Unfortunately, our mother became very ill after the birth of her last child and she was unable to continue working in the garden. I cannot remember the garden being used again during our time on the station and it finally curled up and died from lack of use.

Not far from the main homestead; the station owners before us had built a house which was used by the top stockman and his family.

The station's previous owners had also installed a concrete swimming pool up the far end of the house which we quickly learnt not to use. It was not the most pleasant experience swimming in the pool and turning and finding next to us members of the reptile population swimming as well. So we much preferred to have a swim in the river.

Life on *Hillside Station*

We had moved to *Hillside* in late 1952 if my memory is right. We didn't want to leave *Mt Vernon,* and to make matters worse it was a sheep property. Father did not like sheep all that much so his long term plan was to change to cattle as soon as he could. Father found the change from cattle to sheep quite difficult to adjust to. Mother on the other hand liked father working with sheep and thought it was a great idea as it meant that Father and the workmen would have a safer environment with less accidents (or at least less serious accidents) for Mother to take care of.

Heritage Listing

Since we left *Hillside Station*, it has been given a Heritage Listing.

> *Hillside Station* has aesthetic and historic significance. It represents a period in the evolution of pastoral stations...
>
> Physical Description

Hillside Station lease was taken up by George and John Gregory Withnell in 1880. It was often referred to as '*Shaw River Station*" in early writings. *Hillside* Homestead is situated on the western side of the Shaw River. A small stone house was built and additions in later years have been added. The homestead complex consists of various buildings. Photographs show dome shaped roofs, which are effective against cyclones, and wide timber verandahs supported by timber posts.[47]

.oOo.

---

[47] http://inherit.stateheritage.wa.gov.au/Public/Inventory/Details/5b2182c3-bd66-4f2a-89e8-595221455ea9

CHAPTER 7
The Four Musketeers

Remembering back, as an adult, to our childhood growing up on Mt Vernon Station, we were on the whole a merry little group. There was my eldest sister Maureen, then Jerry and Tadgee our Aboriginal playmates, and our two younger sisters Judy and Elizabeth who were only babies (or in Elizabeth's case not yet born) as well as myself.

As children we had what could loosely be termed a nanny. Fanny was an aboriginal lady who lived on *Mt Vernon*. Fanny and our mother became very good friends. Fanny was like a mother in the way that she watched over and comforted our mother, especially when she first went to *Mt Vernon* as a young bride.

Fanny was also like a second mother to us children. First it was Maureen and I that Fanny took charge of, then Judy and Elizabeth when they were born. We were her family and she was our family as well.

It is mainly Maureen and I who I remember about when the family were living on *Mt Vernon*, as our sisters were still quite young when we left *Mt Vernon* and moved to *Hillside Station*. Growing up on *Mt Vernon* also gave us children the chance to mix with and to have Aboriginal friends to play with and get up to mischief with. Again, that was mine and Maureen's forte.

Aboriginal Style Education

Tadgee and Jerry were our comrades in arms. They taught us to speak their language and to track native animals. They taught us to follow the footprints if we were trying to track an animal or people. All of us children would play trying to track a person or one of us children or an animals, be it a bird or something else. We became quite good at these tasks; so much so that our parents were very proud of their children and their skills! Mother and Father called us their two little angels and loved to show us off to visitors at every opportunity that came their way.

Except we were not the two little angels that they thought we were. When Mother found out from Fanny what we were really saying she was horrified and not at all impressed with our skills after all. Maureen and I were marched off to the bathroom, where Mother then took a bar of soap and proceeded to wash out our mouths. YUK! I think I promised to be good through a mouth full of water and soap. Maureen, on the other hand, told mother the soap tasted very nice, so needless to say she ended up getting her mouth washed out twice. Sometimes it just does not pay to be so smart, as she found out.

With our two friends, Tadgee[48]and Jerry, we were always getting up to mischief of one sort or another.

Trouble and mischief followed us everywhere, so much so that our family claimed it was our middle name. Having the freedom to roam around the homestead grounds as well as the surrounding bush was magic for us children. We would be gone

---

[48] Tadgee was found by Fanny and one of the other women; she had been left under a bush by some Aboriginal people who had been visiting the station. It turned out that Tadgee was not wanted by the mother or the tribe for some reason or other. I am not sure why that was; it may have been because she was a girl or even that she was possibly a twin and her sibling had died, so they looked on Tadgee as bringing bad luck to the group. That would have been taboo. Anyway, Fanny and Mother took her into their care and raised her. Tadgee was an interesting little girl, full of mischief and she was up for anything.

for hours with our Aboriginal friends as they had even more freedom then we did.

The tribal elders were teaching Jerry and Tadgee the Aboriginal skills of their forefathers which they would need for living off the land. During this learning stage, the four of us explored to our hearts' content. We would find ourselves wandering far and wide, building cubby houses and finding treasures. As long as we were home before dark our parents never seemed to worry too much and we always had a house dog with us who would know the way home if we became lost.

The elders taught us what fruits and berries were safe to eat, and showed us what native trees and shrubs we were not to touch in the area; and if Maureen and I were lucky or quick, we ate with the Aboriginal families at lunch time. That is, if we could sneak away from Mother's watchful eye.

Maureen and I were incredibly lucky because the Aboriginal women always included us in those lessons along with Jerry and Tadgee if it was possible. As well as learning what berries and roots were safe to eat, we were shown how to cook goanna and kangaroo and, believe it or not, also snake and other native animals. They also taught us how to find witchetty grubs which were a white little grub that lived in holes in the tree. From what I can remember it was a very tasty little grub.

The four of us would spend hours practising our new found skills. We would walk for miles tracking little or imagined animals after we were taught the art of tracking, following their footprints in the sand. Most importantly, we were shown how to find water in the bush. Bushcraft was – and still is – a very important skill for both the Aboriginals and the white man.

I can remember heading up to the Aboriginal camp just as soon as I had bolted down my boring lunch so that I could have a proper feed of native tucker. I am sure Mother knew what I was doing but I think she left me in Fanny's capable hands to make sure I did not get into too much trouble.

The four of us were always getting up to mischief or, more to the point, getting into trouble. We always managed to find things to do that resulted in a spanking for each of us.

Mischief in the Tomato Patch

One incident was to involve Mother's veggie garden and Tadgee and myself. Mother and Father had built up a kitchen garden so that there would be fresh vegies for all. Until one afternoon when Tadgee and I came up with a novel way of keeping ourselves amused. Playing a made-up game which I will call 'rounders' (the rules were made up as we went along) which meant hitting little tomatoes and running as fast as we could in a circle, back to what we called home base. We had spied the tomato bushes full of green tomatoes and what a grand time we had as we picked as many of the tomatoes from the tomato bushes as we could and then we set up our little rounders circle and what I will call our pitch. The two of us went through a lot of tomatoes. What a great time we were having – until we were spotted by one of the Aboriginal ladies and reported to Mother and Father. I can remember my bottom smarting for a week after that.

What Tadgee and I had thought of as having a great time impacted on the food source for the family and station workers. It meant we were short of tomatoes for that year. It was not till I had grown up that I truly understood what Tadgee and I had done that day. Our reasoning was that because the tomatoes were green they would be no good to eat. How wrong we both were.

Rounding up the Chooks

Maureen, Jerry and again Tadgee also managed to get themselves into trouble big time which on that occasion resulted in a spanking from our father.

One of our much loved games was pretending to be stockmen like our father and his workers. We would climb on our make-

believe horses and round up whatever was available, be it
sisters, playmates, cats, dogs or, best of all, chickens. The pigs
were just a bit too big for our games.

One day Maureen and co (Jerry and Tadgee) were amusing
themselves playing stockmen and their target for mustering
that day was the chickens. First the gate to the hen house had
to be left open. After that was done the chickens were rounded
up and pushed out the gate and into an open area just outside
the chook run. This was when the fun was nearly ready to start.

Climbing aboard their make believe horses and with little stock
whips made by our father in hand, Maureen and Jerry were
ready to start the muster. First they had to make sure that they
could get the chickens back into the chicken run. Once that was
sorted, the yelling and chasing of the chickens would begin.

Being good stockmen just like our father, the two of them were
full of energy and just busting with excitement to begin their
game. The poor chickens were to be shown no mercy by
Maureen or her intrepid stockmen.

There were chickens squawking and flying in every direction.
Up trees, on to the roof of the hen house, even up high on the
fence and anywhere else the chickens thought would be safe
from Maureen and Jerry, with little Tadgee following close
behind.

What the three of them forgot, when they started their muster,
was that our father was home that day and would see and hear
what they were doing from the homestead veranda. The noise
from the poor chooks and the yelling from the three children
soon alerted our father to a possible problem. The next thing
that happened was when our father, with his belt at the ready,
came striding up from the house and towards the hen house.
Needless to say my sister and her friend and fellow stockman
Jerry also had smarting bottoms for a week or so. Tadgee had
managed to hide from Father so escaped the spanking. The

excuse that Maureen and Jerry gave our father was that they were practising to be stockmen, just like Father and his men.

Make Believe Station Management

Maureen was a very feisty little girl and had a very strong survival instinct. This made her the ideal leader of our little team. Maureen took great pride in being our leader, and she was always organizing our escapades. I was mainly a follower and went with whatever the group wanted to do most of the time.

All of our adventures were set around our station life and activities. We acted out make-believe mustering trips with our broom handle horses. As well we acted out branding and butchering our make believe stock. Maureen always played the father role and I was made to be the wife or one of the animals which we had on the station. The same went for Tadgee and Jerry.

One day Maureen was busy running our make believe station and it came time in the game to rub down and feed the horses, milk the cows and chain up the dogs. Unluckily for me, the dogs went bush when they saw Maureen coming their way so I was made to take the place of a dog. Maureen chained me up and went on her way but, unbeknown to her, she had made the collar too tight and I very nearly asphyxiated before one of the women found me and ran for Mother screaming "Virginia's dead, Virginia's dead". From what Father and Mother told us, it was a very close call.

Pay Back Game

There came a day when Maureen decided that it was time to pay me back for cutting her head with an old rusty tin after we had got into a fight.

Maureen being Maureen, there was some planning to be done. First she had to find a suitable place, well away from where our parents could see us. Then the ground was carefully marked out and Maureen told me where I was to stand and where she would stand. There were also ground rules set in place, like how far apart from each other that she would allow us to stand. After all, as Maureen said, the aim was to hit one of us with the stones. Also the size of the stones that she allowed us to throw at one another was important. This episode was carried out with great seriousness.

Jerry was made the referee – much to his horror – as he and Tadgee tried to talk Maureen out of the pay back game (as she called it) but Maureen was not to be stopped. Tadgee even threatened to run and tell our parents on her, but even that did not stop Maureen. The game was to go ahead. Maureen selected the rocks for us to throw, but Jerry stood his ground because he said the rocks were too big for us to use and made her find smaller stones. The result was that Maureen managed to hit me on the head with a small stone and Jerry then declared the game was over.

Doing Housework

When we were very small and Mother and Fanny washed the blankets, we were made to hop in the large washing bowl (which was more like a trough) and they would urge us to jump and splash around till they felt that the blankets were clean. We thought it was great fun as we could make as much of a mess as we liked and we would not get into trouble from Fanny or our mother.

Horse riding

One day Father and the men thought that they would teach us children how to ride. For me it was a bit of a disaster, as I was thrown more times then I stayed on. The horses were not gentle creatures but working horses so were not used to having small children sitting on their backs who could not control them. I don't think Maureen or Tadgee fared any better than I did. Jerry, on the other hand, was a real star; he was born to ride.

The upside to learning to ride meant that the four of us loved to hang around the stockyards and watch while father and the men were breaking in wild horses to use for cattle work. This was a very exciting time on the station calendar.

I remember that our love of hanging out around the stockyards meant that we were always getting in the road. Finally, Father or one of the station hands would manage to find little jobs for our idle hands to do. The four of us just wanted to share in everything that was taking place down at the stockyards.

From the stories that Father and Mother told us children over the years, Maureen and I (as well as Tadgee and Jerry) were very lucky to have reached adulthood because of the mischief that we all got into.

Figure 21: Working horses in the stockyards

One of the stories that Father loved to tell happened when the station menfolk were breaking in wild horses. Father and the men had gone up to the house for a smoko break (morning tea) but they had left a stallion loose in the yard while they were away. Father and the men had been trying to work with this horse before they left.

This stallion was wild and did not like people to boot; all the men including Father had been kicked or tossed by this horse so it was considered dangerous and Father was the only one who continued to work with it.

When Father and the men returned from their smoko, they got quite a fright on reaching the stockyards. They were brought to a sudden halt as they looked on in horror at what was taking place in the yard, or so the story went.

At this time Maureen and I, from what our parents told us, would have been about three and four. Tadgee would have been the same age as me, and Jerry was a year or two older.

The four of us had climbed into the yard where the stallion was, and not only had we done that, but we were climbing all over this very vicious horse. The stallion had one child hanging off each leg, and from time to time we would hang around its neck. And this wild beast that our father and the men had been trying to work with was as gentle as a lamb, just quietly walking around the yard with us all hanging on. From what our father told us, we were having a great old time.

This nearly caused our father to pass out from shock, at the sight of us with this horse. The next problem that Father faced was trying to get us all safely off the horse, and out of the yard.

He had to use all his horse skills (of which he had many) to gentle the animal so it would allow him to remove us children from its back and then to get us out of the stockyard safely. This he did, and our father went on to name this horse Blaze because of the white blaze down its face.

What amazed Father and the men was that Blaze would still only let us children near him for quite some time after that day. But after seeing how Blaze related to us children, and the way that the horse continued to be gentle with us, Father believed that Blaze would be a very good and intelligent horse if he could be trained to work with the cattle and other station horses.

Father's judgement of Blaze was spot on, so Father – who was very skilled at handling horses and cattle – decided to give Blaze a chance and continued to work with him every day. Finally he and Blaze came to an understanding. Blaze went on to become Father's main stock horse, particularly in mustering season.

The amazing thing about this story of a man and his horse was that Father had this stallion literally eating out of his hand. After washing and combing his horse, Father would put out a feeding bag of chaff but he would keep a little of the feed back. Then he would hold out his hand towards Blaze. The horse would first nuzzle our father and then take the feed gently from his hand. Sometimes father would have a carrot or an apple. As children we were lucky to see this comradeship between our father and his horses many times.

Molly, the Shetland Pony

Our father was given a Shetland pony named Molly for us children to ride, but we were only allowed to ride around inside the boundary fence of the homestead because Mother and Father were afraid that the mare would take off if given the chance, into the bush never to been seen again.

Maureen and I as well as Tadgee and Jerry had loads of fun on Molly. Then she became pregnant and after a time became just too big for us to get our little legs around, so father put a stop to our riding till the foal was born. After Molly had her foal, Father or Mother named the foal after our sister, Judy. What a little monster it was; we could not get near Molly without being given a good kick and it would bodily push us away. So no more riding Molly.

CHAPTER 8
Station Life

Father had been on *Mt Vernon* from 1929 to the early 1950s apart from when he was in America. Mother joined him there immediately after their wedding in 1936.

Station life began in the cool of what was commonly known as Piccaninny Dawn, which is just before the sun comes up. In summer this was about four o'clock in the morning. In winter it was a bit later, around five o'clock in the morning. The day did not end till the sun went down in the evening, and that could be any time from six o'clock or seven o'clock at night, depending on the time of the year.

If mustering and droving were taking place it could turn into a twenty four hour day for father and his men as they had to keep the cattle settled or they could stampede if something frightened them, like a dingo or lightning. They could also be away from the homestead for six weeks – some times longer – during this time.

Figure 22: Cattle at a water trough, stockmen on horseback

When father bought the land he made sure that there was enough water so that he could put down wells all around the property. He was careful to build stockyards near where he had sunk these wells. From my memory, there were about three or four stockyards around the station, and always near a water source.

Managing the cattle

Mustering was done on a regular basis to keep track of the cattle and to make sure that the animals were healthy. This was what our father liked to call maintenance mustering. While the cattle were in the stock yards, waiting to be branded and ear marked, Father and his men would check them to make sure there were no other brands or ear marks on them, as that would mean that the cattle that they had mustered were possibly from other stations in the area. If so the cattle then had to be returned to those properties.

Mustering was the time when station owners sorted out the cattle and returned what cattle that did not belong to them to their rightful owners. Cattle duffing was always a problem for station owners – and probably still is – so care had to been taken that the strays did not get mixed up with father's stock.

It was quite common for cattle to stray over boundaries because kangaroos and emus would push their way through the fences, leaving big holes. Needless to say, so would the cattle, especially if they spotted lush feeding grounds on the next door property. This was when diplomacy was needed the most as the stock were sorted and returned to their correct owner. The brands and ear tags were the way that the cattle people could identify which station the cattle had come from.

Twice a year there would be the big muster when Father and the men would be gone for many weeks, depending on what the seasons had been like in terms of rain. There were a number of stockyards scattered around the one million acre property and these yards were used during the big musters.

The musters were also the time that Father would make his selection (this was also known as drafting) of what cattle he would send to the markets down south. They would muster wild cattle to send to the markets in Perth. The cattle, after being mustered, would then have to be ear marked and branded with the owners brand mark. (I cannot remember the brand for either *Mt Vernon* or *Hillside Station*). After being drafted from the mob (that is, from the rest of the cattle) Father would then select the cattle he planned to send to Perth markets

Mustering and droving meant that father and his men would be away for up to six or eight weeks – or longer – depending on where the stock agent that Father was using at the time wanted the cattle to be taken once they arrived in Perth. If it was the Fremantle Port, the cattle would droved to Onslow to be loaded onto a ship. They were then shipped down to stock agents Elder Smith or Dalgetys in Perth. The same method was used when the cattle were to be sent by train; they would be taken to Meekatharra and put on a cattle train and taken to Perth or Midland where they would be held at either Elders or Dalgety's holding yards near Perth.

Mustering and droving

It was in the early 1940s on *Mt Vernon* that our father bought a ute. Up till then, all the station work – mustering and droving – was done using horses, and carts were used to transport the food and the stockmens' swags, as well as whatever else he thought they might need while they were away working. The carts and drays were pulled by big draught horses.

Mustering and droving was done by men on horses. It was only when we were at *Hillside Station* that father started having his cattle trucked out.

On the droves, father took with him two horses for riding plus one horse as a pack horse to carry the swag. There was also a wagon which carried food, water and equipment. Father would usually have four stockmen with him, each with several horses.[49]

One year when I was about two and Maureen was three, we went droving with our mother and father to Cue. Maureen can remember that, and it was reported in the local paper.

> "Mr and Mrs Arthur Meehan and children, of *Mt. Vernon Station*, arrived at Cue safely last week together with a head of cattle. Mr and Mrs Meehan are staying at the Club Hotel, Cue, awaiting trucking facilities for the cattle."[50]

The owners of stations on their way would let the muster drovers feed from their paddocks and also drink from their rivers and waterholes. If they had stock to sell, they might arrange for the drovers to add them to the mob on the move as reported on another occasion.

> *"A mob of 440 cattle reached Onslow from Mt Vernon Station during the week, having travelled approximately 400 miles on the first stage of their journey to the Midlands Junction sale yards. The cattle in charge of Mr Arthur Meehan of Mt Vernon reached Onslow in good condition. Mr. Meehan said on arrival that he was forced by exceptionally dry conditions to thin out his stock, not having had any rain for two years. On the way to town, cattle were picked up from other stations to be shipped away on the Koolinda."[51]*

---

[49] One story that J.P. told was about drovers that were taking a drink from the river when one man yelled "What the bloody hell was that?" It turned out it was a freshwater crocodile. The men moved out of there very fast, so J.P. said.

[50] *Daily Telegraph and North Murchison and Pilbarra Gazette*, 21 September 1945, page 3: Cue Section.

[51] *Northern Times (Carnarvon)*, 9 November 1950, page 3: Onslow Notes

Figure 23: Stock horses wait in the shade at Mt Vernon

In the 1950s trucks were taking over the job of transporting the stock to Perth, and now planes and helicopters as well as motor bikes and quad bikes have taken over. I am not sure if many of the station hands even use horses now. The big change would have to be the satellite phones that all station people use now and then there is the solar panels that provide power to the homestead and outer buildings.

Working the Cattle

Everyone on the station looked forward to the mustering, especially if the cattle were brought into the homestead stockyards which were only a short distance from the house.

Back then, the cattle that father and the men mustered and sent south were what was known as bush cattle. This meant simply that they had never been handled (worked) and were wild and dangerous for those having to work with them.

Father would spend a few weeks each year mustering the bush cattle, and – from memory – he would use the bush cows for breeding purposes. He would put his selected bulls to the bush cows and this way he would always be improving the quality of his beef cattle as well as of his breeding stock that he used on the station. The wild young bulls were castrated so they couldn't breed with the cows.

The branding and ear marking of the cattle as well as the castrating of the young bulls and dehorning were all done in these stockyards. Father would then sort through the cattle to see what he thought would be suitable to send to the markets.

Father and his men used stock horses that had been trained by our father and one of the other stockmen (I cannot remember his name) for working with the cattle, but even with all the care that Father took, things could and did go wrong which in some cases resulted in someone being hurt or put at risk of being hurt.

At *Mt Vernon* we would all line up along the homestead fence, which faced a large stretch of claypan that ran down one side of the homestead. This was ideal as it could be used when bringing the cattle into the stockyards. From there we could all watch as Father and his men drove the cattle past the house and down to the yards.

This was a very difficult and dangerous task for father and his men as the cattle that the men were mustering were not used to being handled. They were very edgy and could be spooked by any unexpected sound or movement. This meant that the men handling these cattle had to be calm and reasonably quiet. Keeping quiet also meant us children watching from the house had to also keep quiet as our yelling could cause the cattle to stampede.

I can still remember one incident which involved Tessa, one of the Aboriginal ladies who lived on the station. This lady had gone out for the afternoon to pick wild berries and to gather firewood, but she did not know that father was bringing in a mob of bush cattle to the stockyards. Father of course had no idea that Tessa was out gathering wood and plants for her evening meal.

Tessa had no way of knowing when Father would be bringing the cattle he had just mustered past the homestead. He used this area when he had finished the main muster and was getting ready to bring the wild cattle into the homestead's stockyards.

As I have said before, father was unaware that Tessa was out there till the cattle became unsettled and then started to stampede. That was when our father saw Tessa caught out in the middle of the claypan with nowhere to go.

By now those of us standing at the fence saw what was happening, so began screaming for Tessa our friend to stand still. Very still. Mother and Fanny were doing their best to keep us calm and to stop us yelling because that was making the cattle more unsettled. Fanny meanwhile was also yelling instructions to Tessa to stay still.

While all this was taking place, Father and the men were trying to stop the stampeding cattle from running amuck and injuring themselves or the stockmen. Then to our father's horror he saw one of the bush bulls heading straight for Tessa. The bull's head was down and it was pawing at the ground and snorting as it was getting ready to charge Tessa. Which could have meant her death.

Father and the men were in a hopeless situation because they were unable to get between the bull and Tessa without hurting her. Meanwhile we were still screaming, and by this time we were all crying as well. Fanny was yelling at her friend to "put the wood on top of your head and stand very, very still, don't breathe, whatever happens don't move or breathe." Thank goodness, Tessa heard Fanny and that is what she did. The bull came right up to her, snorting and pawing at the ground around her legs, but Tessa stood as still as a tree and held her breath.

This must have lasted for seconds, maybe even minutes, before the bull lost interest and walked away. After that Tessa was in no state to move, let alone walk anywhere, so one of the stockmen picked her up and put her on his horse and brought her in for Mother and Fanny to look after.

Stockyards work

Even as young children Jerry, Maureen, Tadgee and I were expected to help around the stockyards after the cattle had been mustered. They were brought into the homestead stockyards and Father would always make sure we were kept fully occupied, and not up to mischief. That way our father and his men could carry on with their work; branding, ear marking and dehorning the cattle, and castrating the young bull calves.

Figure 24: Cattle in the yards

As young children, we were kept quite involved with what was taking place. The jobs our father gave us were varied. We often had to hold the leg ropes that had been attached to the animal's legs (the ropes were used to help keep the animal still), which took all of our strength. Father all the time would be yelling at us to be careful.

We were also given the job of keeping the fire going that was used to heat up the irons, and to carry the very hot branding irons from the fire to where the men were working in the yard. We then had to pass the irons through the wooden rails to Father or one of the men so that they could brand each of the animals with the station's permanent identification mark. This was a set of letters and numbers burnt – if I remember correctly – on to the rump of the animal.[52]

We were also expected to hand Father or his men any tools that they needed when working with the cattle. Sometimes we also helped move the cattle out of the yard and into the nearby paddock. This could be a very scary job, especially if the animals were agitated. It was not uncommon to see one of the men having to jump for his life as one of the bulls tried to have a go at him. Our father also had to jump the fence many times

Looking back with hindsight (or as they say, twenty-twenty vision) this was a very dangerous task for children so young to be involved with. Maureen, Tadgee and I were about five or six at the time. Jerry would have been maybe two years older, but I have to say in defence of our parents that this was a common practice on stations, for the station owner or the manager to allow their children to take part in station work.

As I have mentioned, this type of work was always very dangerous, and how we managed not to fall into the fire I do not know. Jerry was given the more risky tasks because he was a boy and bigger than us girls and that little bit older.

---

52 Father had branding irons made with each of his children's initials on them while we were living at *Mt Vernon Station*. After one of his accidents, Father was left with a scar on his left ankle which was in the shape of a V; our father always laughed and said he now carried my brand.

Even though station work was hard, it was also exciting, and the four of us always looked forward to mustering time. Maureen, Tadgee and I were never allowed to take part in the actual mustering though. Mother and Fanny had put their combined foot down and declared we were far too young and Mother said that type of work was too dangerous for us girls. Jerry was allowed to go with Father and the men a few times and always came back with exciting tales to tell us.

Horses

While living on *Mt Vernon,* Father bred and trained race horses, and as well he bred horses for the Western Australian Police Force.

I can remember two of the Police hierarchy coming to see Father about buying horses while we were down in Perth on holidays one year. Maureen and I were both quite excited seeing these two men all dressed up in uniform standing at our front door looking very stern and asking to see our father. It took some convincing from both Father and the two policemen before we settled down and went off to play. Coming from the station to the city (or the Big Smoke as the city was referred to back then) made us quite curious around strangers, especially men in uniform.

It has been said that father had some luck with the horses that he bred and trained for races. I can remember one Onslow Race Meeting when father won a number of trophies – in 1951 or 1952 I think it was. The horse that he had been asked to work with had been treated very cruelly by the previous trainer down in Perth. When the owner came to father for help, the horse was in a very bad way and no one could go near it; those that tried to ride it were tossed off, kicked and even bitten.

Bucking and biting were the challenges that faced our father when he agreed to help, so down to the beach Father and horse went and into the water; shallow water at first.

Figure 25: Arthur Meehan on horseback at Hillside Station

There the horse bucked and bucked, tossing Father a number of times, so he took the horse further out to where the water was much deeper and the horse had to swim. Up and down Father and horse went, then he took the horse out of the water and on to the sand and began walking the horse up and down the beach. The horse would toss Father off when he tried to climb back on, but our father was not deterred; he just got back on again and again. Back they would go again, first into the shallow water and if the horse bucked Father off there, he would take it further out so that the horse had to swim. This went on for most of the morning.

Finally horse and man became friends and the owner was able to race his horse in the Onslow Races. The horse went on to win the races it was entered into and as a result Father was awarded a trophy for best trainer. He received a double spirits decanter set, a silver cutlery set and a beautiful bottle of Italian Liqueur. These were given pride of place on the china cabinet in the dining room on *Hillside Station* the whole time our parents lived there.

Milking the House Cows

Our parents believed that we had to be taught about station life and what that entailed. They also wanted to find a way of keeping our escapades under control. One way they did that was to give us chores to do around the homestead. This meant washing the dishes after meals, setting the dining room table for meals, and making our own beds and keeping our bedrooms tidy. These tasks were mainly given to Maureen and me.

We were also given the task of milking the house cows, every day. Our father showed us what to do and at first we were always supervised by an adult; either our father or the governess who was working on the station at that time. Maureen loved doing the milking, especially squirting milk at me from the cow's udder. Sometimes our father would let Maureen and I ride one of the cows around the yard. She was one of our favourite cows, very quiet and gentle. We were also allowed to sit on the back of one of the other cows while Father milked her.

Butchering

We were also expected to help our father with the butchering of an animal for the meat which was eaten by all who lived on the station. This would take place about every four weeks or so.

I was given the task of helping our father with the butchering from when I was a young child, both on *Mt Vernon*, and *Hillside Stations*, up to when I left *Hillside Station* in 1959. I suspect that Maureen, Tadgee, and Jerry found more important things to do out in the surrounding bush when Father called for our help, so in the end I became the main helper; more by default than any other reason I think.

Early on the station had no refrigeration so our father built a large cool room not far from the kitchen. This building was quite interesting for that time, as it had a tin roof and the walls were made from two layers of chicken wire mesh. Spinifex was packed between the two layers of mesh. A watering system was then attached to the roof and walls so that when the water was turned on, it would to drip down the walls to keep the spinifex damp and the room cool. It worked a treat. Mother and father used it all the time to store perishables and meat. Once we got a fridge, they used the cool room for home grown vegies and fruit and other things. The spinifex was changed every year.

Teasing the Horse Hair

We also had another job which none of us were all that keen on, but Father liked us to help with teasing of the horses' tail hair. This took place about twice a year, but it was hard on our skin because if we were not careful the horse hair would cut our fingers.

Father would trim the tails of his horses once or twice a year. This was to stop the horses' tails from getting caught on fences or fallen tree branches or scrub when the men were out mustering.

Mother used the horse's tail hair for stitching up wounds suffered by our father and the stockmen during the course of their work on the station.

Mother would put the hair in boiling water for about ten minutes. It would then be hung out to dry, after which it was stored in the medical chest till needed. As children if we needed stitches for a bad cut, mother would use the stored horse tail hair as thread. After I had a bad fall from one of father's horses when the family were staying near Onslow, our mother used the horse hair that was in her medical chest to stitch up a large cut on my face.

Mother, apart from using the horses tail hair to stitch wounds, also used the hair for stuffing mattresses, dining room chairs, and her lounge room couch and lounge chairs when they needed to be restuffed. The horse hair would be replaced about every twelve months or so.

Father too would use the horse hair to repair and stuff his saddles when needed. The leather from slaughtered animals was also used for other jobs around the station. Father made his stockwhips and bridles from the leather. Our kitchens chairs were also covered in cattle hide and they were – from memory – very comfortable.

Our parents made the task of teasing the horse tail hair a family tradition. We would all gather under the shade of the jacaranda tree which was just outside the station's office and radio room.

The hair would be soaked in water over night. This made the hair a lot easier to work with. While we worked our father and mother would fill in the time by telling us stories to make us laugh, and our father would start a sing-along and he always encouraged us to join in. Mother claimed she was tone deaf so would just sit there and enjoy watching us sing along with our father. From memory we had a lot of fun when this happened and the time passed very quickly.

Mother and Father were well aware that teasing was not our most favourite job, so they made quite an effort to keep us interested in other things while we were working with the horse hair. This seemed to make the time go quickly as well.

Our father would often hold us spellbound with stories of America "The land Of The Bold And The Free". We heard these stories many times while we were growing up. I think father would have returned to America to live had he had the chance. I can remember him broaching the subject a number of times over a meal with mother while they were on *Hillside Station,* but our mother had buried two children here in Western Australia and she was not going to leave them for America if she could help it.

Starting the Generator

One of my most disliked jobs was to start the old petrol generator on *Hillside Station* at night when father was away or unwell. The old generator was only ever allowed to be run at night to charge the radio batteries. Maureen and I always got that job and, looking back, I wonder how on earth we did not knock our teeth out because the crank (handle) that we had to use to wind up the motor on the generator would keep coming off and hitting us in the face or the head.

The other worrying job to do with the generator was that Father believed that the generator had to be refilled while the motor was still going, and as the generator got very hot if it had been running for any time there was also the risk of starting a fire. So great care had to be taken by whoever got the job later in the evening to do the refilling of the generator. I so admire my sister Maureen and how she faced everything without flinching for years. I used to have nightmares about life on the station. I am sure life on *Hillside Station* must have stood Maureen in good stead as she now has her own farm. I do not think anything could possibly faze her now.

Sheep work

Maureen and I were recruited by both Father and Mother and put to work. It had become normal for us that when we were home on school holidays we would take on station duties. Around the homestead helping mother with cooking, cleaning and taking care of the garden, and as well as the homestead duties, we would help Father with his stations tasks. On *Hillside Station* we became good as rouseabouts, cleaning and handling the freshly shorn fleece, as well as bringing in the sheep that were to be shorn from the holding yards. Maureen and I quite enjoyed these sheep duties, as we found doing the cleaning and cooking back at the homestead could get a wee bit tiresome, what with mother always running shotgun on us.

Our days helping Father would start from daylight and not end till dusk as we worked around the property together. Maureen and I would help as best we could, with repairing windmills and fixing fences as well as cleaning out drinking troughs for the stock.

What we most liked were the trips back to the homestead after a long day. Father would always start up a sing-along and we would join in. This was a very special time for us with our father as it also meant that for a short time the worries of the world could be put to one side.

Swimming

At *Hillside Station*, if there had been any rain and the river had water in it, Maureen and I could also take our two sisters swimming in the river. That was not always the case, but sometimes there was a good rain and the waterholes were full, or we were able to find a waterhole deep enough to swim in.

Figure 26: Shaw River swimming hole at Hillside Station

Anyway we were happy to give the girls the benefit of our limited swimming skills, which was a form of freestyle made up by Maureen and myself. I must admit it was all a bit hit and miss, but we had a lot of fun. We even took photos of ourselves in our very unfashionable swimsuits which were a mix of dresses, shorts, long pants or if we were lucky a swimsuit – we were a sight to behold I think.

Adventures

When young, Judy and Elizabeth were aspiring little geologists and they were forever collecting pebbles and stones as well as the odd stick or two which they added to their collection. Going into their bedroom could be somewhat interesting as we had to step over all sorts of things which were lying on the floor.

They also liked to take a box of matches and go out for the day making little fires to boil their billy. How they never burnt themselves or fell into the river I do not know.

Social Life on the Stations

The one really big thing for Mother about life on the station was the aloneness. Meg stayed with Mother for the first few weeks when she was first married to help her settle in when she first arrived on the station, and used to come and stay with mother quite often while we lived on *Mt Vernon*. Meg's husband Jack Morrissey would also come and stay. We had few other visitors.

Maureen and myself knew very few children to play with as we all lived far away from each station and would take two days or more to reach each other. Everyone was too far away to go very often so we had no friends other than Tadgee and Jerry.

No children meant no friends or social life for us children. For Maureen and I, it could at times be somewhat hit and miss for us to see other children apart from Tadgee and Jerry. Judy and Liz were only babies then.

Our parents were good friends with the other station families in the district, as Father was a great raconteur and would keep our visitors enthralled for hours with his often very funny stories. Father was a very charming and honest man and people had great respect for him because of that. They also loved Mother for her sense of humour and dry wit. Because of our father's skills in both horsemanship as well as with cattle our neighbours also admired and respected our father, and would come to him for advice. However, distance was always against very much social life.

Station life could be a very lonely and isolating existence for those who lived out on these large properties; so on the rare occasion that they had any visitors they were greeted with great pleasure as it meant that they could all catch up on what was happening in the district.

There were two visitors who stand out in my memory though.

The French Connection

I remember that back in the 1950s we had a visit from two men; this was at *Hillside Station*. The men had seen my father while he was collecting mail from the mail plane. When they heard who he was they arranged to visit us at the homestead. The surprise was that this man's name was Langoulet, which was the name of our father's great grandfather and they were apparently related. I cannot remember his first name. This visitor spoke French and was very fulsome type of person. This was when we found out that our father could speak French too – he had learned it from his grandmother who was the daughter of Louis Langoulant. (See Appendix 2)

Anyway he was a very French type of person and when leaving he kissed my mother's hand and Mother was over the moon. I do not know if he ever came back or if my father was able to track him down. He was a lovely man – possibly he was another great-grandson of Louis Langoulant.

A Cousin Visits

My father's Uncle Will had headed to WA with his brother (our Grandfather), J.P. Meehan. He later returned to Victoria with his wife and family. Then one day his daughter Olivia (our father's cousin) came to visit us on *Hillside Station*.

Olivia had married Les Falconer. Les had been gassed in World War 1 and wasn't well, but after he retired, they travelled round Australia in a VW Kombi Van. Olivia turned up at *Hillside Station* late one day, just as it was getting dark. She had walked for about three of four miles to us to get help from where they had become stuck in the river.

That was how I met Olivia and found out about my cousins. What a wonderful lady. They settled in Perth for a time, and Les died there in 1964. Olivia returned to Melbourne, and years later I lived with her there for a time before I married.

Marble Bar Race Weekend

When we reached our teens, our parents let Maureen and I attend the Marble Bar Race weekend. We were lucky, as we were able to stay with friends of our parents who lived in Marble Bar.

We would go to the Marble Bar Races on Saturday afternoon, and then we would go to the dance, which was held on the Saturday night. Next morning we would all go to the Catholic Mass, which was from memory held in the same hall.

Corroborees

At *Mt Vernon*, Jerry loved it when the men held a corroboree (Native dance of the Australian Aborigines). These were held quite regularly and were not far from the homestead. Our parents were often invited to go and watch.

One of the funny stories that our parents told us when we were growing up was that the Aboriginal men would raid the clothes line looking for brightly coloured clothes to use in their dances.

Clothes that they liked as it turned out were quite often the ladies underwear. Mother would tell us that she and the other ladies working or visiting the station would look up to see their clothes being worn in dances, and quite often on the men's head as a hat. I gather there was much laughter when this happened.

.oOo.

Chapter 9
Mother's Nursing Skills

The work carried out on stations was fraught with danger, and the people who worked on these properties were very aware of the risks that faced them when they left the homestead each day. Accidents happened, no matter how careful our father and his men were.

Our father was well liked and respected by the men who worked for him as he would never ask any of his men to carry out a task that he would not do himself. He was always calm and very patient when working around cattle and horses and this instilled confidence in those that worked with him.

Medical attention on the stations

Father always used to joke that he married Mother because she was a nurse. I think in fact that what he joked about was true as Mother's skills as a nurse were called on many times to help when father or one of the men were injured.

Living on Mt Vernon Station meant that they were well away from hospitals or any other form of medical help when and if an emergency arose. The only way of getting help to the station was by pedal radio, so Mother's skills as a nurse were invaluable to all that lived on Mt Vernon. There was also a well-stocked medical chest provided and stocked by the RFDS.

Isolation was – and still is – the enemy of station people and Father and Mother were very much reliant on themselves. Mother had to take care of many life threatening emergencies while living in the North West.

Father was very aware of the dangers and risks that his family and those that worked for him would face, while Mother on the other hand had to learn on the run so to speak. She faced having to set possible broken legs, head injuries, and many others wounds caused by the cattle, and even snake bites. Once on *Hillside Station* Mother and Maureen had to prepare a body for the RFDS to take back to Port Hedland.

Jack Hammond

While we were living on *Mt Vernon,* Jack Hammond came to
live with us. Jack had been a friend of Father's for many years
and had worked on *Mt Vernon* with father from when the
station was first established. When the 2nd World War
happened, Jack and Father both wanted to enlist in the Army
but father was refused because he bred cattle for the Australian
market; and also because our father broke in wild brumbies
(horses). The Australian Government believed the horses could
be used by the Army and the cattle would be used by both the
army and our own domestic market. This was the case, I
believe, for a lot of the station people and the farming
community during World War II. I understand that the
Australian Government had zoned him and the station as an
essential service.

Jack Hammond did enlist and was sent overseas where he
ended up in Germany, as a Prisoner of War (POW), for a
number of years I believe. When Jack returned to Australia he
was a very ill man and the doctors did not hold out much hope
for him. When Father and Mother heard the news, they both
spoke to the Army and their doctors asking that Jack come back
to *Mt Vernon* to live out his days with the family and the life he
loved. Mother took on the task of bring Jack back to good health
with great gusto so that before long he was his old and happy
self. Jack Hammond was a much respected friend of the family
and lived on *Mt Vernon* till it was sold in 1951. Jack then
retired and moved to Meekatharra I understand.

Dental Practice

Father and the men who worked for him as stockmen had many
accidents – from being thrown from horses to being gouged by
bush cattle – but they also had tooth aches and colds and other
various health issues so Father, being a good and efficient man,
made sure every angle was covered. He also reckoned that by

having Mother by his side, what could possibly go wrong, but they both over-looked human nature itself.

Because of the remoteness of *Mt Vernon* a trip to the doctor or dentist in Meekatharra would take a number of days; this was before an air strip was put in. So Father took it upon himself to learn a bit of dentistry at the Dental Hospital in Perth. Father probably spent a week at the Dental Hospital and was taught mainly how to extract teeth and attend to other minor medical and dental problems that needed attention. When he finished, he was given a case of dental instruments to take back to *Mt Vernon*.

Their planning and preparedness was really quite a sight to behold, until that fateful day when a simple exercise grew less simple by the minute. A day when both Mother and Father's skills were needed, and as luck would have it, needed by one of the station women who came to father complaining about a bad tooth ache. Father set up his dental table and was ready to attend to his patient, an Aboriginal lady of some note and disposition. But trouble soon raised its ugly head and a change of heart was definitely on the cards when this lovely lady spotted father's dental instruments set out on the table. No way, she told our parents, was she going to let him near her, tooth ache or not. No amount of coaxing worked, so what were Father and Mother going to do?

The only way that this problem was going to be solved was if this lady saw Father at work on someone else's mouth first; then and only then would she agree to him treating her. So how to solve the problem? What was to be done and who could they use to demonstrate on? None of the workers who were watching on in great delight were willing to help Father out of his dilemma and offer their teeth!

Father had to find a guinea pig, and quick smart, but who had a loose tooth? Not him or Mother but Father was not discouraged. "Check the children" Father told Mother. Well, Maureen went bush as fast as her little legs would carry her. Me, I was so busy watching all the excitement that I did not hear or see Father coming up behind me; next thing I knew, two big hands were holding onto my shoulders. I was trapped, well and truly. No amount of pleading or struggling worked. I was going to lose a tooth whether I liked it or not.

"Open your mouth, dear" said Father, pushing and wiggling my teeth as he looked for a suitable one to pull. As luck would have it, I did have a slightly loose baby tooth for Father to demonstrate on.

Father explained to me why this was so and that I had to be a good girl and not cry; after all I was a big girl now. Father told me that the lady with the tooth ache would not allow him to remove her offending tooth unless he did it on someone else first. As Father put it, he had to show her that he knew what he was doing on me first. No matter how much I protested.

After all this fuss the lady concerned was more than willing to let Father have a go and after the tooth was removed she was happy to tell everyone it did not hurt, the Boss was a Good Fella.

Only the next day she was back, still complaining about a tooth ache, much to Father and Mother's surprise. "We took out your sore tooth yesterday", said Father, but the lady still insisted she had a sore tooth and began pointing to another tooth. As it happened, the day before she had told Father the wrong tooth was hurting!

"Well," said father, after having taken a good look at the offending tooth from the day before, "It would have had to come out sooner rather than later anyway," so Father told the lady that she was lucky as she had got two offending teeth removed for the price of one.

Mother's Tooth

Mother also had to call on Father's dental skills while they were living on *Mt Vernon*, but for her it turned into quite a drama. Mother was in great pain and her face was all swollen from a back molar that had become infected and would have to be removed. Help was three to four days away and our parents both thought that the problem with her tooth was too serous to wait that long; Father would have to remove the offending tooth. Now, living in the outback there was no such thing as a local anaesthetic available for them to give Mother, so she was going to be in a lot of pain while Father tried to take the tooth out.

There was nothing in the RFDS medical chest that they could use, so they had to come up with a plan. Now both our parents loved a challenge so, given time, they were able to work out a way to remove the tooth and keep Mother from suffering too much.

Mother elected to lie on the bed and our father attached two leather straps to the bed rails at the back of the bed; he told our mother to hold on to the straps on either side while father then started working on removing the said tooth. Maureen and I were of course very interested in what was happening, though I don't think we showed all that much sympathy for our poor Mother's situation; after all, we had been down that road and had Father's skills used on us.

Being a curious little thing of about four or five, I got a wee bit too nosey and wanted to know just how this was all going to work. Much to Mother's horror and Father's amusement, I came up with the classic "I am not going to grow old like you, because you lose your teeth and have to wear glasses when you grow old".

After those rather unforgettable words, Father hurried me away, before Mother could take to me with whatever was close to hand. It was years later before Mother would laugh about it.

Both Father and Mother at that time had to use reading glasses for close up work. When Maureen, Elizabeth and I were in our teens, we also needed to get reading glasses; I think that could be seen as Mother's revenge.

Castor Oil

One of our parents' health prevention ideas was that everyone (and I mean everyone on the station) had to have a dose of castor oil every few weeks. Our parent's thinking was that if we were regular then we would have less health issues. Now this was probably true, but they struck a problem as none of us, children included, would willingly take the castor oil for them. So again father came up with a grand plan, or so he thought.

Maureen and I were made to line up and take the castor oil first; we were then given a slice of orange from mother's fruit trees or if mother had any home-made boiled lollies we were given one to suck on and get rid of the horrible taste. This was the reward we were all given for taking the castor oil. Finally the station folk would line up and be given their dose of castor oil for that month. We were forced to go through this charade every time it came around to taking the castor oil, which would again be given out to us all.

Maureen and I became the station's guinea pigs. When any one had to have medicine or have wounds stitched it was always "Boss! Maureen and Virginia go first then Boss if the kids are OK we will go next". So wait they did and, if all was well, whoever was sick or injured would go after us. That went on till we moved from *Mt Vernon* to *Hillside Station* and were sent away to boarding school.

There were many and varied ways that our parents would use this method to treat everyone on the station. We never saw Mother or Father take the dreaded castor oil; they were just a little bit too clever for that.

Epsom Salts

On *Hillside Station*, Father changed to Epsom salts as the cure-all.

Father would start doing the rounds of the bedrooms at around 5.30 in the morning; incidentally this also happened to be father's favourite time of the day. Father would come to our bedrooms and if we were not up it would be "Wake Up! Wake Up". We could hear father yelling as he came along the veranda, but no matter how much we tried to hide under the blankets it was to no avail, father would still be standing there drink in hand, saying "Drink up" and with no intention of moving till the last of the Epsom salts mixture was gone. Yuk! Once that was drunk he would make us drink a very strong cup of black tea which had lots of sugar in it. Again Yuk!

Figure 27: Windmill, tank and stock trough on Hillside Station

This was our father's favourite exercise every morning when we were home from school on holidays and it worked. We were out of bed and dressed in double quick time before father was back with another cup of black tea. Then Maureen and I would join Father as he worked on fixing fences and windmills on the property and checking to see that the stock had water and feed. Maureen and I enjoyed these tasks as Father was always very appreciative of our help.

Gored by a Bull

Life on the station could change from farce to drama very quickly; maybe the result of someone being thrown from a horse and breaking a leg or, worse, from being horned by one of the bulls while working with them. Then there were the odd car accidents, which seemed to be very rare from what I can remember. Falling down a well while trying to fix it was always a possibility unfortunately, but our father always instilled in his men the need to take care, no matter what the task was.

Our father was pretty good at first aid – on both his horses and on his men – though there was one time it didn't help on *Mt Vernon Station* when father and his men were working at one of the station's stockyards which was quite some distance away from the homestead. They were handling wild bush cattle that needed to be branded and ear marked when a bull broke away from the stockmen holding it and charged one of the stockmen, trying to gore him. The stockman was lucky and managed to climb up and over the high fence.

The bull then turned his attention to our father who was trying to shepherd the other cattle out of the way. The bull saw his chance and went for our father, catching him in the head with its horn. The bull then tossed father up in the air and over the stockyard fence.

Father lay on the ground unconscious for some time as the stockmen tried to work out how to get him home to where Mother could look after him. None of the men knew how to drive. Our father had not at that stage got around to trusting any of them with his newly purchased ute so all they had to work with were their stock horses. They found themselves stranded and waiting for our father to regain consciousness.

They were hoping Father would be able to drive the ute, and get them all back safe and sound to the homestead.

It had become a matter of when or even if Father would come to in the minds of the stockmen. Meanwhile, as they waited for him to come around, they emptied their water bags over Father's head in the vain hope that it would wake him up. This all took time, but finally Father 'came to' and was able to drive every one back home.

By the time they arrived back at the homestead, he was covered in blood from his head injury. Maureen and I were the first to see our father because as soon as we heard his ute, we had run outside to welcome him home. We were stopped in our tracks at seeing Father all bloodied and being helped by the men to walk. I headed for the bush and I think Maureen was hard on my heels. We were both terrified and thought our father was going to die.

Mother stayed calm and went about setting up her nurse's equipment. First she cleaned the blood from his face and head and then stitched up the large gash left by the rogue bull. Father was then put to bed and Mother kept a very close eye on him over the next week or so. She was also in constant contact by Pedal Radio with the doctors at the Meekatharra Hospital during this time.

Father should by rights have been in hospital, but as there was no airstrip on Mt Vernon at that time, (only the large claypan area about a mile from the homestead, which the RFDS pilot was not prepared to take the risk of landing on) Father had to rely on our mother for his medical treatment. She was the only help available.

Day to Day Injuries
Writing about our father's close brush with death still makes the hairs on the back of my neck stand up. Maureen and I were quite young when this all happened, but we were there and we

lived with the fear of losing our father. I am the only one who has kept those memories I think, but for me that memory, like others in this book, has stayed with me for life.

With Father and his men there were always other more minor injuries that mother was able to take care of. Then of course there were us children who also managed to get sick or suffer an injury to ourselves, most of which Mother could look after. At various times our mother had stitched up some injury or other that I had inflicted on myself and I am sure Maureen, Judy and Elizabeth also had mother sewing up some injury on one or the other of them as well.

Elizabeth went on to more dangerous things after we went back to school in Perth. One day she decided to go for a ride on Father's horse which he had left tethered to a horse rail out the front of the homestead. The problem with that was the horse was only used to our father riding him, so that ended with Elizabeth having a very nasty fall and landing on her head and becoming concussed.

Medical Evacuations

At separate times, both Elizabeth and Judy had to be flown to Perth or Meekatharra hospitals because they were very ill. With Judy it was tonsillitis and with Elizabeth it was for a chest infection.

Judy was born on the 14th August 1947. When she was only a few months old, she became very ill with tonsillitis. Mother thought Judy was going to die if they could not get help for her. Our parents – mainly Mother – were on the Pedal Radio morning and night trying to get help and Judy was becoming sicker by the day.

Because they had no airstrip when Judy became ill the family could not expect any help to come from the RFDS. The pilots were too nervous to even try landing their plane on the large claypan strip of land near the homestead.

This was a very trying time for mother, but fortunately for her a doctor living in Perth who was a Ham radio enthusiast had heard Mother's call for medical help and flew his little plane – not sure what type it was – up to *Mt Vernon* to see if he could help. Dr Harold Dicks landed on the claypan stretch of land without any trouble. After treating Judy he then flew both Judy and Mother back to Perth where Judy was admitted to the St John of God hospital in Subiaco in Subiaco. Judy ended up there for about three weeks, I think it was.

While Judy was in hospital, Mother stayed in Perth to be with her and had booked into the Palace Hotel while she waited. Doctor Dick's wife, Nancy, felt that Mother should be with friends, so she asked her to come and stay with her until Judy was better and ready to go home. Mother and Nancy stayed friends for the rest of their lives.

It was after Elizabeth was born in 1949 and we had all returned to *Mt Vernon* that Father arranged with the Country Roads Board and the RFDS to put in an airstrip. This was most fortunate as later on that year Elizabeth had to be rushed to hospital with a severe chest infection. This time, mother was able use the RFDS to take Elizabeth to hospital in Meekatharra.

Sewing Machine Fingers

Not only was Mother having to attend to the injuries of our father and his workers, she also had to watch over us children. One episode involved me and my interest in Mother's Singer sewing machine. An opportunity came about one afternoon when she was having a rest. I must have been about four or five at the time. Anyway I thought it was too good an opportunity to miss, so sat down at the machine with my little piece of material and away I went. Things were going swimmingly till I managed to slide my finger under the needle as I was trying to push the material through and of course the needle went right through the material and my finger and out the other side.

The next few minutes were filled with fear as well as pain as I faced the fact that I was in trouble – big trouble – with Mother, and it was my fault entirely because I had disobeyed my mother and played on her sewing machine. So I sat there, filled with guilt, and with the tears running down my face, trying to work out how to avoid being caught by our mother and receiving a belting from her as well, once she had managed to remove the sewing machine needle from my finger. So if you can imagine for a moment my predicament here, I was desperately trying to work out how to remove the needle from my finger before she found it and me. What was I to do?

I was too scared to call for help, so I just sat there till one of the station ladies found me and went and called my mother. So, after giving me a well-earned telling off, mother went and got a pair of pliers and pulled the needle from my finger. Ouch, that hurt! Mother showed me no sympathy and I was told it was my entire fault for not doing as I was told.

Years later, Mother was relating this story to my two younger sisters while telling them not to play with her sewing machine. But Elizabeth did not quite believe what Mother had told them so decided to check it out for herself. Off she went and following carefully Mother's story, pushed her finger under the sewing machine needle and then pushed down on the foot pedal, and – yes – she also got the sewing machine needle caught in her finger. I don't think Mother quite knew what to say when she saw what Elizabeth had done.

Mother, it would seem, had to come to our rescue on quite a few occasions as we were growing up. If it was not Maureen and I, it was Judy and Elizabeth who were just as mischievous as we were while we were young.

Allergy Problems

An incident that took place near Onslow which involved our Father was about 1945. I find that I have very little memory of this accident, so this story will be very vague I am afraid.

Our father was helping out a fellow cattle man by mustering his cattle ready to go on the ship to Perth when Father had an accident when he and his horse crashed into a low wire fence. This fence was not visible because of the bush growing in the area so was not seen by any of the stockmen on their horses.

Father and his horse both fell badly which resulted in father sustaining injuries and his horse a broken leg. The horse had to be put down(shot). Father was taken to the Onslow Hospital first, then airlifted to the Port Hedland Hospital where he received emergency treatment.

The hospital contacted Mother who was in Perth. Maureen and I were going to St Joseph's school in Subiaco for a few weeks at the time. On arriving home from school, our mother told us what had happened.

While our father was in the Port Hedland Hospital he was treated by a doctor who forgot to check and see if father had any allergies. The doctor gave him penicillin but Father was allergic to penicillin and had even told the medical staff that. Because of the injection Father became very ill and there were complications. He had to be flown to Perth by the RFDS for treatment. As I said early on, I have very little memory of all the drama. Maureen told me most of this story.

The Good Shepherd Home for Wayward Girls

Mother was in fact an amazing lady in many ways and to help her with the running of the homestead and looking after us children she started to employ young girls from the Catholic Welfare group "The Good Shepherd Home for Wayward Girls". These girls in fact were children brought up in Catholic orphanages and, while growing up, they were trained as house maids or office workers or even sent into factories to work. The children were also often sent out to farms or pastoral properties to work when they were about fourteen or fifteen years old.

The orphaned children stayed at the orphanage till they turned sixteen I think it was, then they were normally sent out into the world to look after themselves. But there were some young girls who did not conform to the standard set by the people running these orphanages, so they were sent to the" Good Shepherd Home" which was more like a work house that the girls had to work and also live in. Not a very nice place at all and the girls were treated very badly by the so-called caring staff.

We had a couple of these girls come and live with us on the station over the years and they all turned out to be wonderful people. That is, till they sent mother a fifteen year old girl who I will call Angela. Angela was a very emotional girl and the poor lass had been treated very harshly by the staff at The Good Shepherd home which left her very defensive around strangers and any one in authority; this included our mother and father.

Maureen and I loved Angela and she was very kind to us while she lived on *Mt Vernon*. Unfortunately, Angela was an orphan and being at the Good Shepherd Home was very damaging for her, and she was left with problems that our mother was unable to help her with.

Angela suffered from really bad mood swings; these became worse during certain times of the month, and she would fly off the handle at any little slight, real or imagined, without warning. Mother could normally calm her down, but sometimes this was not possible and things would get out of control very fast and become very frightening for Mother and us children.

One of these episodes took place while Father and the men were out working. Angela became very upset over something – not sure what – and pulled a knife on mother, threatening to kill her and us children. Mother tried to calm Angela down but nothing worked. On this day Angela just became more agitated and violent. After some time, and many more threats, the young lass ran away from the homestead and into the bush.

Father and the men searched for Angela well into the night, but they found no sign of the girl. As well there was no sign of her having returned to the house, so father called in the police from Meekatharra to help find her before she became lost and died, as she had taken no water or food with her when she ran away.

The search went on for about two or three days. Luckily for Angela it was winter time so the nights were cold but the days were not hot when this all took place. There was a great deal of activity going on at the homestead, with the police and their trackers.

The police set up spot lights in the garden and moved all our beds onto the lawn.

The police were worried that Angela might carry out her threat to harm us. I remember us children being in awe of what was happening and we watched the police with our eyes open wide.

We were frightened that the police might hurt Angela, but we were even more intrigued with the carry-on that was taking place around us. We had never seen so many people at any one time at our home. In the end it was father and his men who found Angela; it turned out she had been hiding not that far from the homestead and had found herself a nice little hiding place with lots of trees and scrub to hide in.

When she heard father and the men calling her name she climbed up into the tree canopy and hoped that she had not been seen. The police took her back to Meekatharra with them and we were later told that Angela had been returned to the "The Good Shepherd Home" and that was the one place that she was so frightened of being sent back to.

.oOo.

CHAPTER 10
Watch Over Your Children

As children growing up on *Mt Vernon* and *Hillside* stations, it was in many ways an idyllic life. We were given the freedom to wander off and do what we liked from dawn to dusk, just as long as we returned to the house for meals. To an outsider it would have appeared to be an amazing life that we lead as children.

Speaking only of my own experience, and not those of my sisters, this freedom that had so freely been given and which lead to us children having many happy and exciting adventures also had a very dark side which has had a devastating effect on my life.

Below are memories only of mine. Our parents, on reflection, were oblivious to the dangers that came with some of these men who found their way to Father's properties. Such freedom would in the end become a threat to the safety and well-being of their children because of the type of men our father employed to work on his stations.

To be fair to our parents, they had no way of checking out these men. They either turned up asking for work or were sent by Elder Smith or Dalgetys to the station to work as stockmen. This was the method used by all the station owners needing workers during that time.

In many ways our parents were quite naïve about the evil that some of those men presented to their children. Looking back, I believe that many of these men already had a history of raping and molesting children long before they found their way to *Mt Vernon* and *Hillside Station* where they were employed by our father.

Mt Vernon and *Hillside* stations must have seemed like an open invitation to these men who did make their way to the North West and to the stations there. Some of the men that did come, came either because of a failed marriage or because they were hiding from the law and they wanted to disappear from sight.

Our father, as far as I can recall, never asked any of these men questions about where they came from and what they were doing before they came to the station. My parents' motto was 'do not ask personal questions and do not pry into other people's business'. For my mother that would have been quite normal because of her father and his background.

The isolation and wilderness made this part of Western Australia very attractive for some people and it was not long before they discovered that station owners, and their other workers, asked very few questions. A lot of men found that they could leave their former lives behind and disappear. Other men would re-invent themselves and start new lives with new names and histories.

Mother and Father, sadly for us, were oblivious to the very existence of people such as paedophiles. Lack of knowledge of such people led our parents to not think to provide protection for their children or the children of people who worked and lived on the station.

From 1946 to 1951 a number of men worked as stockmen for my father on *Mt Vernon Station*. Again, when he owned *Hillside Station*. Father also had men working for him between 1952 and 1963.

I was about five and a half (this was about 1946) when I was first raped. This continued to happen for about three years till the men left the station in about 1950 just before we moved on.

I am going to refer to these men only by a number. These men will only be numbered and they will all be referred to as a number.

(1) Number 1 was a vicious and brutal rapist. He would wait for me in the scrub near the chook run or near the pig sty where the pigs were kept till I came past; then he would rape me while

his offsider mate – who would stand watching him – would make sure that no one would see what he was doing. This would happen once to three times a week. Each time I was threatened to keep quiet or certain things would happen to me and my family, they would die. He drummed this into my head to make sure that in my five year old head that I would be blamed by my mother and father and that it was my fault, that this was all my fault and I would be punished by my father and my mother.

(2) This man was different. He was kind and gentle and appeared to care for me. I was five and a half when he started to pay me some attention. He would invite me into his room and I would be given a sweet to suck on. This man would tell me stories or read to me. As well he gave me little gifts, and he made me feel cared for with gentle hugs and kisses – nothing to frighten a small child with. Our parents were not ones to hug or kiss Maureen or myself from memory. My parents seemed not to notice his growing interest in me or how much attention he was giving me. This was when the nightmares and sleep walking and talking in my sleep started as well as bed wetting. The bed wetting stopped when we were sent away to boarding school in Perth but the nightmares and talking or yelling in my sleep have continued all my life.

This attention went on for a number of weeks and during that time he never did anything that was in any way sexual towards me from what I can member, but things started to change one afternoon when he went from cuddles and kissing my neck to touching my body in a very sexual manner which left me (a small child) frightened and confused but again he reassured me with hugs and kisses as he wiped away my tears.

He would also speak gently to me when I became upset by his touching; then he started to bring out his penis and have me touch it. He named his penis Big Boy and would say that Big Boy was now my pony. He would then move me up and down on his lap and say that Big Boy was giving me a ride. This become very sexual and made me frightened and I cried. I did not like having a ride on Big Boy.

One afternoon, after cuddles and touching, he raped me. He then cleaned me up and took me back to the house. This went on for about three years and each time he raped me he gave me a gift. One day he gave me a doll; he would give me the doll to hold when I cried or whimpered during the rapes. The doll was always kept in the man's room.

Judy was only a few months old when she became very ill with tonsillitis and had to be sent to St John of God Hospital in Subiaco in Perth. Judy and mother were away for close to four weeks and during that time Maureen and I were left in the care of Fanny and our father. It was during the time that Mother and Judy were away that the attacks on me from two of the men escalated.

As a child I grew to hate dolls. One year when travelling back to *Mt Vernon* from Perth with my father he stopped at a store in Meekatharra and told me to pick out four dolls for us girls. I told him I did not want a doll and that I hated dolls but my father still bought a doll for each of us. It was not long before I had pulled the doll's head off and tossed it away. I told my father that the dog had eaten it.

(3) The third man knew what was happening to me in his mate's room and up near the pig pen and chook house but only ever raped me once. He preferred to watch his mate raping me.

(4) The fourth man also only ever touched me once. This again happened while Mother and Judy were in Perth. One day, after seeing the second man leave the station in the buggy to go and pick paddy melons and to go for a swim, and noticing that he had taken me along (after getting permission from my father to take me) this man became curious and decided to follow and try to see what was going on. What he saw was a common event, when I went with number (2) to the river.

After playing in the water the man would then wrap me in a towel; we were both naked and then he would start cuddling and touching me in personal places till finally he raped me. This happened more than once. We would then have lunch and after lunch he would again rape me, then we would go for a swim and go home.

Once when we had been in Perth, I returned to the station with my father and there was only the two of us there for about two or three weeks. The rest of the family were to return a bit later. It was during those three weeks that the attacks on me again escalated.

When these men returned after working away from the homestead for weeks – or even for a day – they would lay in wait knowing that sooner or later I would have to make the trek up to feed the chooks and the pigs.

Each time that I was sent to do my tasks away from the safety of the homestead, my parents were immune to my tears and would force me to take the frightening walk. My parents just thought I was a lazy child trying to avoid doing my jobs, so ignored my tears. Every time this happened, I found myself trapped by these men and subjected to their sexual assault. One of the men (1) had a way of treating me after each assault – he made me take my underpants and hide them where they would not be found.

Each assault was followed by threats against my family and against me. I was about five and a half when this all started. It was not till the governess came, which was some time in 1949, from memory, that these attacks happened less often.

I was also sexually assaulted by one of my father's stockmen on *Hillside Station* in the 1950s.

It seemed that these men were more than willing to put up with the harsh conditions of station life – in fact they may even have felt it was a very small price to pay – if it meant that they could get their hands on children. Apparently they had no fear or respect for the law or the law's ability to apprehend them and make them accountable for the crimes that they were committing.

Because more than one man carried out these attacks on me at any given time, they seemed to have no fear that our parents would be told about the attacks by the other stockman working on the station.

The Aftermath

In early January of 1949 mother came down to Perth to await the birth of our sister Elizabeth. Maureen would have been seven and a half, I was six and a half, and Judy would have been about eighteen or nineteen months old.

Elder Smith (the Pastoral Company) had arranged for mother to rent a flat in Cottesloe which faced the beach. We could walk across the road, which was not very busy and play in the water or amongst the sand dunes. There was a low brick wall that ran along in front of the sand dunes to stop some of the sand from building up on the road.

The owners of the flat that mother rented also lived there and the lady worked at an old peoples' home which, from memory, was just a short distance away. They also had a daughter Alexandra (Sandy) about Maureen's and my age so we had someone to hang out with. It was a lovely spot for children with the beach just over the road.

Mother was finding these last few months in her pregnancy very hard and would get upset at anything and did not like us children playing outside. I knew this, but asked mother if I could go outside anyway, so mother turned her anger on me.

I had asked mother if I could go and play and that set Mother off. Mother was mad at me. After looking at me for some seconds, mother grabbed me by my shoulders and I was thrown into an over-stuffed arm chair where mother stood over me yelling and demanding why I had asked to go outside. I had no answer to her question.

Soon after we arrived in Perth, Maureen and I were admitted to hospital to have our tonsils removed. Maureen had been having some trouble with her tonsils; so Dr Harold Dicks (a friend and the family doctor) thought it would be prudent to operate on us both. Not that I had had any problems with my tonsils but because we lived such an isolated life with no doctors available if they were needed.

Doctor Dicks had a passion for planes which included a Tiger Moth I think it was and had flown up to rescue Judy when she was a baby.

Mother used to laugh about the plane ride back to Perth, as Doctor Dicks had to land in some farmer's paddock and explain to a rather confused owner that he needed to refuel his plane so that he could continue his return trip to Perth.

The same situation arose when Elizabeth became ill as a baby, but by then Father had built an airstrip and the RFDS were able to come out to the station and take Elizabeth away to the Meekatharra Hospital for treatment. So Doctor Dick's and Mother's concerns were well placed. This was in 1949.

The hospital where we had the operation was in Subiaco. The Matron there was an old friend of our mother so we were very well looked after and even given a surprise gift of ten shillings when Mother came to take us home.

With two children being operated on in hospital and a little one with her in the flat, Mother was starting to become quite stressed at this stage. Also she was getting closer to when her baby was due as well.

Our Aunties (Mother's sisters) were very good and tried to make things a little easier for mother, but tensions arose between her and Maureen. This was not that unusual as both were very strong willed people.

Things reached boiling point one day between Maureen and Mother. I have no idea to this day what triggered the confrontation between them.

Not long after we arrived in Perth Maureen and mother had a hell of a fight. I'm not sure why. Maureen found herself trapped in one of those big over-stuffed armchairs and mother was yelling and hitting her. Maureen had her little legs drawn up under her as she tried unsuccessfully to make herself as small as possible and by doing that, get out of reach of mother's hands to some degree. We would never know what had started the trouble between my sister and Mother.

Finally it became too much for me as I stood trembling and waiting for things to escalate to the point where Maureen or even Mother could be hurt.

I said it.

The silence was deafening and it seemed like the World must have stopped for a few seconds.

Mother slowly turned with a shocked look on her face. "What did you just say Virginia?" Being so young I did not realise that I needed to be very careful about what I was going to say next, so I dutifully repeated.

"Will you F.....g shut up"

Mother slowly closed her open mouth and silence settled on the room; a silence filled with fear and trepidation. Mother still had her hand raised in mid-air as she had been about to hit Maureen when I spoke. Now that hand was turned on me.

After looking at me for some seconds, Mother grabbed me by my shoulders and I was forcibly thrown into one of the over-stuffed arm chairs where mother stood over me yelling and demanding answers to her questions.

With her hands now on my shoulders, Mother was shaking and yelling at me while I sat trembling in the chair, trying to make myself as small as possible so that I would hopefully disappear into the chair out of her angry reach.

Maureen still sat huddled in her chair with a look of shocked surprise on her face as our mother had stopped yelling and hitting her; Maureen was now watching spellbound as our mother turned the full force of her fury on me

"From the men who worked for Daddy, on the station when they did those horrible things to me," I tried to tell her.

Mother, after taking all that in, wanted to know more. When it had happened and who had done it? The whole time she was shaking me and yelling that it was the most disgusting thing she had ever heard, and I was a filthy little girl.

I was still hoping that now that I had been able to finally tell her what had been done to me by those men on the station I would be safe because my mother and father would protect me from those men and she would love and keep me safe.

What a shock I was in for as the tirade continued; I was called a dirty disgusting little girl who was to blame for what had happened to her.

Worst of all my mother told me that she did not want me to touch her – 'never ever' were her words – and finally, did I understand that I would for ever be 'soiled goods'.

The world as I had known it came to an end on that fateful day. I had disgraced my family, lost my mother and father and my sisters in one dreadful afternoon.

I withdrew into myself and in many ways became a loner; I learnt to keep my thoughts to myself and not make any demands on either my father or my mother.

After that dreadful day, I found myself turning more and more to the beach across the road from the flats. I would go there first thing in the morning and play and hide in the sand dunes. I only came back to the flat for evening meals. It was a dark, dark world for a little girl to find herself in.

Sometimes I would go for a swim, but I had no idea how to swim, so I believe that the gods were watching out for me as I always seemed to find the strength to dog paddle back to shore after I had floated some way out of my depth and was heading out to sea.

Mother's sister Meg, after hearing about that explosive afternoon and what had happened to me and seeing our mother's reaction, took me back to her farm for a couple of weeks. Years later she talked of this very sad little girl who had withdrawn and never smiled any more.

Before mother went into hospital to have Elizabeth she had arranged for a babysitter to come and stay with us children in the flat while she was in hospital. Elizabeth was born on the 7th March 1949 at St John of God Hospital in Subiaco in Subiaco.

We returned to *Mt Vernon* about a month after Elizabeth was born. Aunty Meg came as far as Geraldton on the train if I remember rightly, and father picked us all up from the Meekatharra railway station and drove us back to *Mt Vernon*.

Mother never spoke about what had taken place between her and me while we were in Perth, but she went to a lot of trouble to avoid my touching her till the day she died. My father also avoided touching me whenever he could.

Back at *Mt Vernon,* the assaults went on and each time I would clean myself and then dig a hole in the dirt under my mother's orange trees or lemon trees. I was told by one of these men I must hide my underpants in the most unlikely place. Mother had grown orange and lemon trees in the garden next to the bathroom and I dug a hole and buried my underpants and covered them up. I dug a hole and hid my underpants. It was very important to me that Mother or Fanny never said anything to these men. I was so frightened of these men. I had managed to dig little holes just under the trees and no one asked why. Mother looked at me and thought I was playing a game I think.

This little story sounds quite odd now but I did my orders with great care and spent quite some time digging these little holes and hiding my little underpants.

One day I heard mother speaking to Fanny, and it was about my underpants. Mother was saying to Fanny that it was strange that my underpants seemed to be missing. I had done as I was being told to do but mother had no idea what was happening. She just took no notice of me. This did not bother me, and she said I would get over whatever was happening and just thought it was strange, about my missing underpants I think mother just went and ordered more underpants from Elder Smith, most likely, I guess.

Fanny kept an eye on me just to find out what was going on. Fanny thought something was wrong and wanted to help as we were running out of my underpants.

One day I looked up to see Fanny next to me with a worried look on her face. Fanny and I had this conversation that went on between the two of us with many puzzled looks.

"Virginia, what are you doing, making all these strange little holes?"

"It is a secret. You must never tell anyone. Promise?"

"Yes, but why?" Fanny asked me.

"I cannot tell anyone. They will hurt me and Mother, so do not tell anyone, will you?"

Fanny gave me a hug and said "Can I help, Virginia?"

After looking at Fanny for a while and Fanny nodding she promised, the two of us dug up all the underpants that I had been digging holes for.

"We must not tell anyone" I said. Fanny looked at me for some time and then went and I stopped digging these little holes. She washed my very strange underpants and to my knowledge never told anyone.

I think my mother was just very pleased because the underpants started to come back.

After we returned to the station and the assaults started up all over again, I became terrified to leave the house. Today I cannot leave my house even when I am safe. I do not know if my mother ever told Father what I had told her when we were in Perth. I was told by my mother never to speak to anyone about what had happened to me on *Mt Vernon* ever again to anyone.

.oOo.

CHAPTER 11
The Penny Jar

One of the problems of living in the outback was the lack of shopping opportunities. Mail was delivered every four to six weeks; sometimes longer, depending on the weather. Father would often have a long wait before receiving his supplies which were needed on the station. When mother wanted to order Christmas or birthday presents for the family, she would have to place her written order about six months early just to be sure that they would arrive on time. Often even that did not work and they would arrive late for Christmas or our birthdays. I think that was when Maureen and I worked out that there was no Father Christmas; we would have been about four and five.

Station deliveries

To have supplies delivered to the station, Mother and Father would either send a telegram to Elder Smith in Meekatharra or Perth, or write away to department stores in Perth, like Ahern's or Boans. Mother was the one who organised for presents to be sent from the Perth stores.

Supplies for the station were also ordered this way and delivered with the mail, except for large deliveries. When our father needed petrol or kerosene (which came in forty-four gallon drums) this came in later times by mail truck, but earlier on it was brought by camel or bullock drawn drays.

Large or heavy equipment was also brought by camel or bullock teams. Father had to contact Elders and organise for the goods to be delivered. In the 1940s I can remember a lovely Afghan man with a team of camels. I think there were six camels to a team. I cannot remember the name of this man, but he used to deliver to *Mt Vernon* about every six months. He would stay for a couple of days and always gave Maureen and myself a treat. Once he gave us a lovely china tea set.

Figure 28: Camel team delivering stores to Mt Vernon

I think this man lived with his family near Meekatharra. The camels were used to pull the heavy loads. I can remember the camels spitting at Maureen and myself if we came too close, but we were not deterred by that. We wanted a closer look at these strange animals, so we did our best to get up as close as we could.

Father also used a man from Marble Bar; I think his name was Mr Faber. He had a team of bullocks to pull his large dray. Both men used these big drays for hauling heavy or large items. From what I can remember, each dray had a team of six.

This improved greatly with the arrival of trucks, and then planes in 1949. I can remember the camel teams that came to *Mt Vernon*, as well as the early bullock teams. By the time we left *Mt Vernon*, father had had an airstrip built and the station was starting to have regular mail and station supplies delivered. Father also had electric power installed using a wind generator and finally we had power for the radio and house.

Mother's Treasures

Mother had some lovely bone china as well as silverware. They were wedding gifts. Mother would bring these out on special occasions, such as birthdays or anniversaries, as well as when there were visitors.

Mother treasured her beautiful china and would polish her silverware – including her silver teapot and sugar bowl – on a regular basis. Maureen and I were encouraged to polish the silverware as well when we were older and Mother even taught the Aboriginal ladies how to look after her treasures.

Life on *Mt Vernon* went along swimmingly most of the time, but as in life, there are always hiccoughs along the way. This said, Mother's silver teapots (and she had two or three as well as her beautiful china) often met a much messier end. One teapot melted after being left on the top of a very hot stove, and the china just slipped from our fingers (including from Mother's fingers at various times). This would make our mother very sad.

Over time this meant that Mother's treasures – the lovely china and silver tea pots – were decreasing faster than Mother would have liked. As there were no shops available for Mother to replace her goods, she came up with her grand plan. Just like her mother before her would do.

Mother's plan was to have a Penny Jar. This was used as a means of teaching all of us to be more careful about handling Mother's crockery and glassware as well as her silverware. When one of us broke, melted or otherwise damaged Mother's treasurers, we would have to put a penny in the Penny Jar.

The Penny Jar idea came quite some time after Maureen found herself in trouble with our father for breaking a cup and saucer. We would have been about three and four; maybe even younger. Wanting to be good kids and to help, our mother gave us things to take across to the dining room so that she could set the table for the evening meal.

The dining room was set back from the kitchen for safety reasons and also because of the heat that could be generated from the two wood-burning stoves. To reach the dining room you had to walk along a path from the kitchen. This led to the veranda, where there was a small step. At the time Maureen was carrying a cup I think and I was given spoons to take across to the dining room. Anyway, our father was telling us to hurry along. Maureen was trying to obey our father and hurried just like he had told us, except that when she reached the veranda and the small step she lost her balance and fell, which resulted in the cup being broken and our father becoming very cross and yelling. Hence the Penny Jar.

Mother's Penny Jar method – to our father's horror – also included him and any of his workers who happened by chance to damage any of Mother's treasures. (Though Father from memory never broke any of Mother's precious china ware). Our mother was very pleased with her grand plan, and kept an eagle eye on the Penny Jar as well as us to make sure we were putting pennies in the jar.

The Penny Jar was a success for a while, till mother realised that none of us had any more pennies to put in her jar. Mother was the only one on the station who had any pennies and as *Mt Vernon* was out in the middle of nowhere, once Mother's pennies ran out that was the end of Mother's plan. Mother never did find a solution to saving her precious china from being broken so she would only allow them to be used on very special occasions.

Mother managed to get her plan to work for a very short space of time which included the horror of Fanny breaking a precious cup and saucer of Mother's. Mother was not a happy camper and told Fanny that she would also have to find a penny to put in the Penny Jar.

Fanny just gave a loud laugh, looked at our mother and, after some deep thought, said with a smile "in your dreams". It was not long after that talk with Fanny that Mother removed her Penny Jar from the kitchen.

.oOo.

CHAPTER 12
Kitty's Plans

Mother, for as long as I can remember, always had a plan on the go. Be it having Maureen and I teaching my younger sisters how to ride a horse, or school or swimming or even teaching the sisters to ride a bike.

Arthur remembered what Kitty's mother had said when they were married. "Just keep making plans, Kitty" said her mother. "An Irish lassie girl never stops making plans."

As far back that I can remember my mother was always making plans.

There were plans to change the way the medical trunk was always kept locked. (I think that was Mother's nursing background.)

Then there were the plans for the garden if only it rained.

Mother's big plan was to get as far away from the station as possible. Mother had grand plans of buying a farm or a business down south where Father did not have to work as hard and they would not have any large debts. It was a shame that Father did not listen to his wife!

As long as I can remember, Mother was always coming up with plans. It could be another plan this week or next year. Mother spend many days coming up with plans. Father played along with Mother but some of her ideas were good and others not so good. One was that we had to teach my younger sisters to ride a horse. Not so lucky.

It did not help when Mother thought we should teach out little sisters to swim. None of us could swim – just dog paddle – but Mother thought it was a good idea. Then Mother thought that riding their bike was a good idea so there we were trying to teach Elizabeth how to ride her bike. That was more or less fair. I think Elizabeth gave up on us, but not Mother. Then I was to teach my younger sisters how to dance and that seemed to work quite well. There was always some idea in our mother's head. Some worked. Some were not so good.

From memory we mostly grew up with Mother's plans. Then we all got married and none of us were into plans. At least I do not think we were.

Educating the Girls

One of the down-sides to living on a cattle station was the isolation from other children. This meant that there was no school for the children to go to. For Mother, living on the stations, particularly on *Mt Vernon,* meant that trying to educate her children was not easy. There were three ways to educate her children.

One was to send the children away to Perth or Geraldton to boarding school, or for the mother to move down in school terms while the children attended local schools in the nearest big town.

Another one was that used more often, which was to try Correspondence (it was called Long Distance Education). Correspondence was not that good because it meant that the mother did all the teaching of her children.

The final idea was to hire a girl to act as a governess.

Mother tried all of the methods in turn to try to get us some education.

Correspondence Lessons

The Education Department ran correspondence lessons for country kids. The correspondence lessons were meant to be supervised by either a parent or in some cases a governess. This lead to classes being missed or poorly supervised by Mother.

Education for children like myself and Maureen was very intermittent; most of the time there was Long Distance correspondence as it was for most of the families living on stations in the North West during the 1940s and the 1950s. It was not until we had an air strip built in 1949 that we had a fortnightly mail delivery to *Mt Vernon* by plane. Till then our mail was delivered about every six months or so.

The school work and work books came from the Perth Education Department and they would send work from Perth for us to do about every six months, from memory. There was also a teacher that visited all the stations who were doing Long Distance Education to make sure we were learning.

Our Governess

Having a governess worked but was expensive. It was what our mother did. When mother came home from hospital after having Elizabeth in 1949, she had decided that because Maureen and I had major gaps in our education (we were 7 and 8 then) we should be taught by Correspondence, and mother hired a governess – Ada Broadly – who came from a farming family down south. Maureen and I loved having Ada live with us and we got to do fun things as well.

The Education Department agreed. Mother could do that after checking Ada to make sure she would be suitable to teach us. The Education Department also made plans to make sure that they sent the school work that we would use. The idea was that Ada would help us to learn to read but that only worked if she could find us before we disappeared – that was Maureen and myself – because we were quick on our feet. Not having any young neighbours near our age meant that we did not see any children to keep our interest to learn alive. But we were caught by a visiting teacher from Perth who claimed that I had a photographic memory which was a very bad thing, at least for me.

Education was very much a hit and miss thing for Maureen and myself as everyone on the station had better things to do.

Whether we actually learned anything before being sent away to boarding school at Loreto in Perth is questionable.

We had very little contact with other children living on nearby stations although they had the same problem. The responsibility rested with mothers to teach them, or hire a governess. When they thought the children were old enough to be sent away to boarding schools in Perth or Geraldton, the children were then sent away. The children today have what is called Long Distance Education (Correspondence) as well but they also have computers and better radios for the children and teachers to keep in touch. Mothers still are the main helpers for their children though, I am told.

Attending School Intermittently

Maureen and I had a number of on again, off again, attempts at school.

My mother took Maureen and I down to her sisters Bridie and Meg one year and while down with her sisters, she put Maureen into the local Catholic Primary school for about five weeks.

Then Mother thought she had a brain wave which she thought would work, so she waved me off to some farmers that Mother said she knew. I am not sure who they were but I stayed with these farmers for a number of weeks, crying all the time, till the man said he had enough of this little girl and he was taking me back to my mother. My mother was spitting chips and kept saying that I was to have stayed with these farmers.

My two aunties made Maureen and I dress-up clothes for Maureen to wear to a school affair before we left as they had promised.

Having small children seemed to make things harder for our mother to take care of Maureen and myself, and if her sisters were not able to take Mother and Maureen and myself in, we were dumped with whoever would take us and God help us. If that failed, we were left at a school in Geraldton. We were twice left at this boarding school in Geraldton. Mother just left us, and it seemed like weeks before we saw her again.

That was our very first experience of boarding school, when we were placed in a Catholic boarding school (at Stella Marris in Geraldton) for a number of weeks. We were about four and three years of age. It did not make a lot of sense at the time but the experience was quite traumatic for us both. The sisters were very kind to me and did their best help me when I was homesick; but the teaching nuns scared the life out of me.

We were again placed in Stella Marris for a number of weeks when mother was waiting to have Judy. Maureen would have been about five and a half, and I was about four and a half.

When our parents had their house in Daglish (a Perth Suburb), one year when were down from the station on holidays we were sent to the Catholic Primary in Subiaco (St Joseph's) which was about four miles away. To get to the school we had to walk through the West Australian Railways Marshalling yards at the Subiaco train station, both morning and afternoon. It was like playing dodgem cars with the trains. The men working there were very good. They kept an eye out for us every school day. Maureen and I went to that school for about six weeks from memory. It was near Christmas one year, and not the best experience from memory.

I liked one of the nuns very much at St Joseph's as she had been very kind to me while I was in her class, so I asked my father if I could buy her a Christmas present, which I did. I bought her a tub of face powder to use, which I then presented with great pride to the nun. I think my parents could hear the nuns laughing all the way back at our house.

Then there was one stage when mother had this brain wave which was to send us to Loreto Convent for a couple of weeks. Maureen and I would have been about nine and ten years of age then I think.

Mother was told about Loreto and I think that was why she thought it might be a good idea. Our Aunty Grace had been a student there when she was young, and so had her cousin Bill Meehan's daughters.[53] It was talked about as the best ladies school in Perth I think. Mother liked Loreto so much, she booked us in as weekly boarders for a month or so. That worked well – or so mother thought – so that before we could say "Jack Robinson" she had us booked in for the rest of our school lives.

Boarding School

Having us at Loreto meant that mother had a reason to come down to Perth more often to see her children. The problem with that was that each time Mother found it very hard going back to the station. Mother was still having moments when she was emotionally fragile and seemed unable to care for her children.

The Loreto nuns decided because of our age we should be put into age-appropriate classes; the fact that we had had no formal education at all mattered little to the nuns. I asked for help a number of times from the teaching nuns but was told not to bother them.

I know I was all at sea in the classroom at Loreto. Maureen seemed to handle it better than me. The nuns at Loreto kept telling me I was stupid but then again so did our mother. The nuns were not interested in teaching me. Take it from me; if the nuns were not interested, you were sent to the back of the classroom. I was humiliated and ridiculed by a very unpleasant nun (as well as some of the other nuns) even when I had grasped the subject and was starting to show improvement in all my subjects.

The nun that ran the junior school was a very nasty person who took great delight in targeting me for her abuse. A day never

---

[53] Bill Meehan, the son of J.P.'s brother Jim, had five daughters who were sent to Loreto when they lived near Geraldton. Bill was our father's cousin.

passed without me being on the receiving end of her vicious attacks, yet my mother thought the woman was a saint. One of the other nuns targeted Maureen, I believe.

Country kids were fair game at that school from what I saw during my time there. A strange thing though was that in the junior school there was a tall and very kind teacher who kept asking me to sing all the time. If she could keep the children acting or singing, then she would have me up singing; not sure why. But I liked that nun very much and she made being at that school at least bearable.

Mother and Father never to my knowledge questioned or challenged the nuns about what was happening. Maureen appeared to do somewhat better and no one ever said she was stupid. I think she was in fact very bright and able to keep up with the school work. I gave up in the end because it was just too painful. The effort to try was a nightmare.

When Maureen reached her Junior year, things changed for her. Instead of studying for her exams that year, the nuns started getting Maureen to be their messenger girl. It was "Maureen do this", "Maureen do that". They tried to do the same with me and I can remember telling Mother Rosalie that, no matter what, I wanted to at least try and do my Junior year.

School holidays

Mother quite liked having us children for the school holidays, mainly because she would have spent hours having a great time coming up with ways and means of keeping my sisters involved – which would really mean Maureen or me. Now Maureen and I are two of six children, Maureen being the elder. Mother had lost two of her children (Mary and the last baby was John Patrick).

Both our parents liked keeping us busy when we were home on the station. We were given house duties and cooking. Maureen was a better cook than I was. I liked cooking main dishes and

Maureen liked cooking cakes and deserts, which she was very good at. We also helped our father around the station helping fix fences and windmills.

Maureen and I liked helping both Father and Mother when we were on holidays. Mother kept Maureen and me doing all sort of different jobs, including helping Father with slaughtering the cattle. This was both for the family meat and also for the man from Port Hedland. I think his name might have been Horse. Not sure but he was a lovely man from my memory.

Both Father and Mother kept us girls busy. When Mother came up with her brain storm, I doubt that she even talked it over with Father before she put her plan into action. Mother had a reputation with her family and amongst friends at the hospitals where she had worked as a nurse which was that our mother (Kitty) loved to plan. And Mother's plans worked most times.

It was the school holidays. (It would have been 1955 or 1956 from memory.) We were told by our mother about her plan on a Saturday. The school had finished on the Friday and we flew up to the Shaw River Mines site on Saturday. Mother called Maureen and myself before she dropped her bomb and Elizabeth was ordered to come to Mother's meeting. Maureen and I thought all would pass so we went with Father to some faraway place, hoping by the time we returned that whatever was brewing at home would be over and we would all have managed to miss the fight or whatever it was. It was not a good idea to come home (even our father knew when to avoid home at any cost), because we knew something was up by the racket that Elizabeth was making. Judy and my little sister Elizabeth were both crying and Mother had a face like thunder and was getting madder by the minute.

Mother was planning to send away my little sister Elizabeth but she was trying to work out how to break the news to us all, but most importantly to Elizabeth. Mother tended to make plans quietly and secretly. Father was never told till after

Mother's plans fell into place. When Mother made her mind up about sending Elizabeth away she had just turned seven in March; far too young to be sent away. Judy was too young when Judy was sent away the year before. Judy was about six and a half then from memory.

We could hear Elizabeth coming from the cool house, which was where we went when we were trying to get away from the heat. Mother was there with my two young sisters. Just as we came onto the veranda we could hear Elizabeth – my sister – who was really kicking up a racket. Elizabeth was as mad as could be, so we wanted to know what was wrong. Elizabeth kept shouting that there was no way was she going to do school work when her sisters would be on school holidays. No way was Elizabeth doing that, that would be most unfair and Judy joined in and supported Elizabeth as well.

Judy was sitting in the corner watching what was going on and crying. Judy always started crying about a week before we left for school in the hope that Mother would fall for her effort and change her mind about sending Judy away to school. Sometimes it worked. Mostly it did not work.

Father finally quieted every one down. "What is going on?" yelled father, and he also told everyone to quieten up as well. Mother pretended that all was OK, but Judy spoke up and told us what was really going on. Mother had just told Elizabeth that she was being sent away to Loreto Convent in Perth at the end of the holidays.

"No." Elizabeth said she would not leave *Hillside Station*. "No way ever", and our father agreed, so what was Mother going to do now to send her away from *Hillside Station*. You could guess what Mothers next plan would be! Mother looked at Maureen, but Maureen said "No!" Not her, so then mother looked at me and then came her next plan. I was going to take over teaching Elizabeth.

Elizabeth would have to do some catching up for when she went to school. Mother was not all that interested in doing Correspondence (or Long Distance). She thought it was a total waste of time so she never really bothered. Father and Mother had sent three daughters away to Loreto to board, but for some reason my father thought that sending Elizabeth away at her age was not right, and he also thought that the cost was too much.

Mother was trying to tell Elizabeth that she had two weeks to catch up on school work because she was already behind, and that I was going to teach her while she was on a holiday.

That was the first that I had heard of this new plan of Mother's. I told my father that was not a good idea and it would not work out for my sister.

Father said "Stop worrying. Remember how you started reading? Well, do the same with Elizabeth. It will work."

Thank goodness that Elizabeth was a quick learner and also had a good memory. At the end of two weeks Elizabeth was reading quite well. What a magic effort we put in. No help from Mother. Just myself and a brilliant effort by Elizabeth.

Father was very angry about the way mother had tricked him and us children, as we found out that mother had been writing to Loreto and had made arrangements to send Elizabeth to Loreto at the end of the school holidays. Father was beside himself, but Mother got her own way yet again. It was about two weeks later and Elizabeth was with us when we all were on board the plane heading to Perth.

Mother had done the same things to Maureen and Judy as well as myself. Mother was very underhand in sending Judy away the year before when she was only six and a half, and Judy was just too young; she never settled down at Loreto. So that was not a very good idea.

From memory, Judy went away to school when she was six and a half. Elizabeth went away to Loreto when she was seven I think it was. When Maureen and I were home on school holidays we had to teach poor Elizabeth reading and maths as well as helping our father with his jobs as Elizabeth was little prepared for formal education. It was just as well that she was a bright child. Father did not believe in school for girls so mother had quite a battle to send us way to Loreto. It is a pity that she was misled into believing that Loreto was a good school.

.oOo.

CHAPTER 13
Snakes and All Things Creepy

This chapter is about the snakes and all things creepy that one encounters every day if living in the North West of Western Australia.

Mt Vernon and *Hillside* stations enjoyed and respected the many and varied snakes, scorpions and other creepy and crawly visitors that made their way to and from our house.

It was also wise when wanting a drink of water to not get into the habit of drinking straight from the house tap or even the garden tap or hose; always let the water run for a while so that the mice or little frogs cooling themselves in the pipe did not end up in your mouth first.

Both stations had a range of snakes, spiders, scorpions and other creepy crawlies so this chapter is going to be about them.

At first Mother was not very brave when it came to snakes, scorpions or anything that was remotely creepy crawly, but by the time she moved down to Perth to live she had managed to overcome most of her fears about snakes and those other nasties.

Ants

We had a variety of ants. Small sugar ants that found their way into everything, and as well there were the meat ants (I think they that is what they were called). These were to be avoided at all costs.

Never find yourself sitting on a bull ant's nest when out helping around the station either.

Termites

Termites were dreaded by all station owners as they had a huge appetite when it came to the house and outer buildings.

Termites were known to eat doors, cupboards and the legs off dining room tables and chairs. It could become so bad that the dining chairs and table legs had to sit in cans of sump oil on both of my parents' stations.

If I remember correctly the bathroom cupboard in Mother's bathroom came close to falling off the wall because the termites had eaten most of the back off it.

Spiders

The poor spiders never stood a chance with our mother; out came the broom and down came the spiders and their webs.

You had to be on the lookout for red-back spiders as they had the habit of sitting under the toilet seat in wait for the next person to sit down. It was at your own peril if you sat down without looking first around the room and then looking under the toilet seat.

Red-backs were found in and under cracks. They were also found under kitchen chairs and stools. White-tail spiders could be seen around the station and you had to be watchful of them. There were also little money spiders that our mother loved and tried to protect from being hurt. We also had daddy long legs and huntsmen spiders; the huntsman spiders could grow quite large and from memory seemed to live mainly around the homesteads.

There was also a spider that liked to spin its web around the veranda posts; it was quite pretty. I am not sure what it was called as I only have a vague memory of this spider.

Bats

There were also little bats that flew around at night after dark; mother loved them. As a child I found them a bit scary because they always seemed to be going to settle in my hair.

Scorpions

Scorpions were a different matter as they interfered with the preparation of meals at night. As soon as the house generator was started up the scorpions came out. They were attracted to the lights of the house, but they had also discovered a tantalising smell that was coming from the direction of the kitchen where lights could be seen. So this was where they were heading when they came out at night. As much as the lights attracted them, they were more intent on reaching the kitchen where the cooking smells were coming from. At night, after the lights came on and the day's heat had cooled, the scorpions started heading for our kitchen; there they would gather right in front of the wood stove. This made it nearly impossible to attend to what was cooking on the stove.

The scorpions stayed in the kitchen till the homestead generator was turned off; around nine or ten at night. Once the lights were off the scorpions would head back to their holes or under their rocks to escape the heat till the next night.

Because the scorpions were quite aggressive, cooking meals often became a battle royal; scorpion verses humans. The scorpions would rise up on their back legs and lunge at us, and if we were to slow to move we would be stung on whatever part the scorpions could reach.

Mother set up a defence against the scorpions which was a 4 gallon drum filled with hot water which was left on the wood stove during the day to heat. There was a jug nearby which she would grab and fill from the 4 gallon drum and then throw the hot water in the general direction of the scorpions.

But we learnt that the scorpions were clever little critters and after a while they started to bring along their little buddies; this way they would line up in front of the stove and us and the battle would be on as to who was going to move faster; them or us. Mind you, at times it was touch and go.

The hot water treatment worked most of the time. I was the one in the family most likely to cave in to my fear of being stung by the wretched scorpions, so I came up with my own method for outsmarting the beasts. I would grab the nearest chair to stand on, and stay there till the scorpions gave up and went away. I also made sure that I had a broom handy to wack them with.

Snakes

Snakes were plentiful on both stations, so our father kept a length of rubber hose hanging outside the kitchen door. If a snake was around Father or Mother could grab the hose and with luck dispatch the snake (which was not a protected species back then).

One morning – much to our father's horror – while he was having his early morning shower he looked down and found himself sharing the shower with a snake. Father did not notice the snake till it slithered over his feet and then he retreated with some haste. After which he declared that the snake could take its shower alone. This happened twice to my knowledge.

Mother had a number of confrontations with snakes as well. Late one night when she went to attend to Judy or Elizabeth when they were little, mother thought she had heard a sound. On looking down she found she had nearly stepped on a snake that was slithering its way under the girls' cots.

Once when father and his men were away mustering and Mother was on her own at the station she had a very unnerving experience with a snake when we were all back at school in Perth. During the night mother had got up to go to the bathroom and was returning to bed. This time she had to deal with a death adder that had become quite agitated at finding itself on an unfamiliar surface. It looked like the snake had found its way up from the garden to the veranda. Finding itself in unknown territory, the snake was ready to attack.

As I have said, Mother was alone on the station when she found herself having to do battle with the death adder. The only thing she had with which to fend off the snake was a broken handle off a shovel.

The battle between Mother and the snake lasted quite some time, according to Mother. It most likely was not that long but felt like hours to her at the time.

The snake was very aggressive and came at Mother time after time. It kept trying to gain leverage by backing up against the nearest wall and then trying to lunge at mother from that position. From what our mother said, it seemed to spring through the air as it tried to attack her. The poor snake was probably just as frightened as Mother was.

Coming face to face with a snake while having a shower or bath (or worse, curled up in the toilet bowl as it tried to escape the heat), was a horrifying moment for the person wanting to use the toilet.

This happened to our father one day when he went to use the toilet next to the generator room. I could hear father yelling for me to come quickly. When I got to the toilet there was our father white of face holding a long blacksmith pair of tongs. He explained to me that a snake was in the toilet bowl and he needed me to catch it and hold it with the tongs and he would try to hit the snake and kill it. So father and I set about our plan; me holding the snake with the tongs and father trying to hit it with his tongs. Well, we lost the snake; it had managed to slide away from me and my tongs. Father yelled that I was not holding on hard enough to the snake and it managed to slither away.

As a small child I can remember our father and his men trying to remove a carpet snake that had made its home in a pea vine that grew on the wall just outside the main bathroom. Mother and Fanny had seen the carpet snake while giving us children a bath. Father wanted to remove the snake but not hurt it. It took some time before the snake could be encouraged to move, then father grabbed it and put it in a bag which the men then took out into the bush and let go.

That carpet snake was one lucky snake as most found their way into the cooking pot of the Aboriginals; they thought a bit of snake cooked in ashes was a great treat.

Father came back from checking windmills one day and he had something in a hessian bag. He told me it was a baby kangaroo. He then asked me to hold the bag, which did not feel right so I said "I am not going to hold the bag."

"What is really in the bag?" I asked my father?

"A large snake!" he replied.

After much laughter from Father and Mother and my two younger sisters he let what he had claimed to be a baby kangaroo onto the ground. It was not a kangaroo, but a very large – and I mean very large – snake. I was then ordered to hold the head end of the snake with the blacksmith tongs while our father measured the length of the snake and it was large. I cannot remember quite how long it was, but the Aboriginals had a very nice supper that night. Father thought it was very funny telling me to hold the snake.

The story was that father had been working down a well, fixing something, when this large snake slithered around his legs. Lucky for Father he had a bag with him so the snake was quickly put into the bag and came home with him.

A favourite place for snakes was the bathrooms. I remember a couple of times going into the bathroom to wash my face and reaching for a towel when I heard a hiss and then a snake slithering out from under the towel rail ready to strike. Another time I was just going to get under the shower when I reached out to put my change of clothes on the kindling box and this very big – and I mean big – brown snake unwound itself and, hissing, it got ready to strike. Needless to say, I took off and called Maureen. We both went to get help from Tommy and Lina. Tommy had Maureen and I looking all around the bathroom and outside in the long grass to no avail. Well, the only help Tommy and Lina gave us was to laugh, because the snake had by then been long gone, probably in the grass outside the bathroom.

One time Mother found a snake in the bathroom after having had a bath. She was drying herself when a snaked slithered out from behind the bathroom cupboard which was on the wall near the bathroom door. Mother yelled for help and our father and his men came running. Father had his trusty revolver and started shooting at the cupboard. Mother looked down to see – no clothes! Poor mother, she forgot to put her clothes on in all the excitement going on after seeing the snake.

Despite the creepy crawlies, life was pretty good for a while and every one enjoyed living on *Hillside Station* and taking part in station life, but that was about to change for us all.

.oOo.

CHAPTER 14
The Loss of a Child

Mother and father were faced with a number of dramatic events during their lives together, but Mother faced two such episodes on her own that changed her life for ever. These are my recollections of that period in our lives.

Their first child, Mary Philomena, was born in 1938. Mary lived only a short time after her birth and Mother rarely spoke about it, but when she did, there was a lot of pain in her voice.

In 1954 Mother was to go through this all over again when we were on *Hillside Station*.

In 1953, mother became pregnant. Both Father and Mother were thrilled to bits. This was what they were hoping for and as well they had their fingers crossed that Mother would have a son to carry on Father's name. Father was not that impressed about having four girls. He thought girls were a waste of time and money – this was said frequently but with tongue in cheek.

I have tried below to write down my memories of what took place on *Hillside Station* that day and the effect it has had on all our lives, over the months – and even years – that were to follow.

Things went well for a while and Mother appeared to be enjoying her pregnancy. Then she developed toxaemia and her doctor – Harold Dicks – ordered mother to Perth for treatment and to be close to St John of God Hospital in Subiaco in case she went into early labour. He was concerned about Mother and her baby as she had developed the same health problems with her five other pregnancies. He believed her to be at risk with this child.

He advised against Mother going back to *Hillside Station* until after the baby was born but Father became very impatient and felt that Mother should be back on the station with him. Also he felt it was costing too much money to have her and the younger children in Perth; and he told Dr Dicks of his concerns.

Doctor Dicks would only agree to Mother's return if she came back to Perth before the baby's due date and that she rested in bed. Mother agreed to all of this, so Doctor Dicks organised for a nurse to return to the station with our mother.

Because a nurse was going to be with Mother on the station, everyone thought that they had prepared for all possible events. The nurse was there to make sure Mother did as the doctor had told her. Mother and the nurse returned to *Hillside* around Christmas time in 1953, and things then went along fairly well till 11th May 1954.

Having a nurse with her on the station seemed to help Mother relax, and it also had a calming effect on our father as well. He did not seem to be worrying quite so much about Mother and the coming birth of their baby.

The nurse made sure that Mother rested, as the doctor had requested. Mother was supposed to be confined to her bed until she returned to Perth, closer to the date that the baby was due.

Mother and the nurse also kept in touch with Mother's doctor, Harold Dicks, as well as the RFDS in Port Hedland.

The events that follow all took place during the May school holidays when Maureen and I were home from boarding school at Loreto Convent in Perth. Maureen had a friend from school staying with us. Mary had never been on a station before, and because she found everything very exciting, it became an adventure for all of us.

The morning of the 11th May started out well, with a lovely blue sky, and just the hint of clouds. The air was crisp but not cold so that meant that Maureen, Mary and I would be able to go swimming later on in the day after we had finished our tasks around the house. So we were all set to have a great day.

Mother seemed to be fine and anyway the nurse was there to take care of her.

Lunch time came and went as normal. Mother went to have a rest and Maureen, Mary and I were getting ready to go swimming in the Shaw River which was a short distance from the homestead. We had not even left the house for our swim when things changed. It would have been about 2pm in the afternoon when we became aware that something was wrong with our mother.

Mother goes into labour

I had gone to get a towel from the bathroom and found our mother. It became clear that there was something wrong. I asked what the matter was, and could I help, as I could see mother was bent over and appeared in pain. After asking again if I could help; mother yelled at me to run and get the nurse, quickly. It was then that I noticed blood running down her legs, so I took off calling for the nurse and Maureen as I went.

By the time everyone had responded to my frantic calls for help and arrived at the bathroom, we found Mother had managed to get herself up to the bedroom and on to her bed. By now Mother had started to bleed more heavily. The nurse and Mother realised that she had gone into premature labour, which was what Dr Dicks had been trying to avoid.

Elizabeth remembers the nurse running around with her hands in the air yelling "Don't die Mrs Meehan, don't die". Elizabeth would have been about five and she stills remembers that day.

As the day progressed it became clear that we were on our own with Mother and the fast approaching birth of her baby.

Mother was becoming very worried about being alone with just the nurse and us five children, ranging in age from five, seven, eleven, and two twelve year olds, all girls. Mother knew that somehow she was going to have to take care of us all.

Living on a pastoral property in the Pilbara meant that you did not have easy access to doctors or medical help when there was an emergency; on this particular day we were at home with Mother and the nurse; our father was away for the day. Now it must be remembered that isolation was, and still is, a big problem for those living on stations in the outback of Western Australia.

This was all taking place in the mid-1950s, when station folk did not have their own planes and were reliant on short wave radio for their communications with the outside world, so there was no way we were going to be able to get help to mother quickly. Because of the station's isolation, the only way we could alert anyone to our situation was by shortwave radio, and this included the RFDS. Satellite phones had not been invented then I believe. Or we didn't have one, anyway.

Mother went into labour early on the afternoon of 11th May 1954. As a nurse, she was very much aware of what she was facing. Poor Mother, she must have felt so afraid, and here she was confronted with the fact that she had to get herself and her children through this. It was going to be up to her. Turning to the nurse and us children who were hovering nearby, Mother did her best to reassure us that everything would be alright.

Mother faced the next few hours with great courage. After going into premature labour at home, mother found herself having to cope with a panicked nurse and five terrified children.

By this time mother realised that she would have to take control. So she quickly worked out who was going to be capable of helping her through the difficult hours that lay ahead for her and her baby.

Maureen was the oldest of us children and the last one to run and hide. So Mary and I were given the task of keeping Judy and Elizabeth out of the way, as we had no idea what to do and were utterly terrified for our mother.

We also had a hysterical nurse to deal with, who just wanted to run as far away as possible, and a dear old man, Charlie (a friend of Father's who had been living at the homestead). The old man, I suspect, was just as frightened as us children.

Mother had tasked me to try and contact the RFDS and tell them what was happening and to tell them that she needed their assistance urgently at the station.

Unknown to our parents, the RFDS head office had made some operational changes with their emergency call signal sometime earlier but unfortunately it had a number of problems so at times did not work as it was supposed to.

Before the RFDS made their operational changes, there was an alarm which the station people or anyone who had a short wave radio could activate when there was an emergency and they needed help.

Those calls were always put straight through to the doctors' living quarters if the office was closed. This was unfortunate for the doctors off duty as it meant that they were always on call for emergencies. That was, till the RFDS changed things and brought in a new method which was to be mainly used when the doctor's office was closed for the day or the doctors were away from the base. If I remember correctly, before the change any calls would be transferred through to the doctors living quarters, but for some reason RFDS decided that the doctors were not to be disturbed when they were off duty.

After my many failed attempt at trying to reach the RFDS in Port Hedland, Mother realised that we were going to be on our own till father arrived home. But I still kept trying, and Charlie kept trying too.

On the day that mother went into labour, Father was far away from the homestead with his friend Bob Otway; they were busy checking fences and the windmills around the property.

Mother was only too aware of the situation that she now found herself in; she was going to have to rely on the frightened nurse pulling herself together and overcoming her nerves, as well as her young children to help her during the coming hours.

After I had settled Judy and Elizabeth down and put Mary in charge of caring for them, Mother put Maureen and me to work as her emergency team. The nurse was supposed to be in charge, and we also had old Charlie. He was called in to help with raising the foot of Mother's bed in an effort to stop or slow down the bleeding. The nurse started to panic when she saw mother was bleeding, and appeared to be unable to carry out even the smallest task calmly, so Maureen took over under Mother's directions.

Needless to say, seeing the nurse in this state did nothing to help us children. We tried to stay calm like mother kept telling us, but as things became worse we could not stop the over-whelming sense of helplessness which was quickly overtaking us. We were in such a state at seeing Mother in this situation we became convinced that she would die before our father returned home.

All we wanted was for our father to get home, because we had such faith in his ability to make things right but as we had no way of contacting him. We quickly realised that we were very much on our own and all we could do was try to keep Mother and hopefully the baby alive till help arrived.

As the evening was growing darker and our father had still not returned home, Mother was the one who remained calm and told us what we needed to do. Mary and I did our best to keep Judy and Elizabeth – who were by now crying and wanted their father – out of the way and with Charlie's help we gave them their evening meal and put them to bed. We were only able to tell Judy and Elizabeth that Mother was having her baby and that she needed them to be good girls and not go near mother's bedroom till we told them it was OK. Also with Charlie's help we were able to start the generator so that we had electricity for the house and to keep the radio batteries charged.

The nurse was in quite a state and kept on saying that she did not know what to do to help mother, and that she was afraid Mother would die before the night was over. The nurse also kept telling me that there was no God, because if there was a God he would never have let this happen to our mother.

Thank God for mother's love of planning, as within minutes of taking charge she had us somewhat organised. Maureen was given the task of nurse as Mother knew that Maureen (like her) loved nothing better than playing nurse, while even though I helped father with the slaughtering, I really hated the sight of blood. Mother knew I would be of little use around her, so she found other things for me to do.

With Mother giving Maureen instructions, preparations were made for her baby's birth. Mother instructed Maureen to stay calm, and to collect towels and sheets and anything else that could be used to stem the blood, which Maureen did. She also did her best to comfort Mother and to help her with the contractions when they came. The haemorrhaging appeared to be getting worse, which had our mother very concerned.

Mother was aware of the state that the nurse was in, so did her best to settle her down so that she would get on with her job of helping Mother with her labour.

There is no doubt in my mind that without Maureen's help, Mother would never have survived that night.

In the middle of all this chaos; we had the nurse yelling at both Maureen and myself and anyone else who was within hearing. I got a special blast for not being able to reach the RFDS in Port Hedland by radio to get help.

Still trying to remember the events of that day; I believe that Johnno Johnson (a tin miner who had set up a mining operation about ten miles from the homestead) dropped in on his way home from Port Hedland to see how Mother was and found us all in a state of crisis. He hurried back to the mine[54] and returned with one of his men (David, I think the man's name was) and they tried to work out what to do.

Johnno and David both tried to raise Port Hedland on the radio, but again with no success

Johnno was aware of the type of station that our father ran and knew that he was probably out working on the property and would not be home till dark. Johnno, David and Charlie gathered around the radio in the office to try and work out their next move. Johnno's man (I will call him David) suggested that he drive into Marble Bar to the small rural hospital that was there and see if they could raise the RFDS in Port Hedland.

Johnno waited around till our father and Bob Otway arrived home, while David set off to drive through the night to Marble Bar; by this time it was pitch-dark.

The road between *Hillside* and Marble Bar was more of a bush track then a road, and like all roads on the station had been formed just by use over many years. The mining companies at that time were using the same bush tracks as the station folk. It was later that the mine management decided to grade the bush tracks that they were using.

Because the roads between Marble Bar and *Hillside Station* were dirt tracks, more or less, it was always best to travel during daylight hours. But David did not have a choice as he had to get help as quickly as possible, so this meant travelling cautiously but with some speed.

---

[54] Johnno, his wife and workers were at this stage living near the mine site. They were using water from the Shaw River to wash or sluice the tin so that it could be transported to Port Hedland by truck and from there sent by ship to Perth.

After arriving at the hospital some hours later and explaining to the staff what was happening on *Hillside Station* David and the matron tried to raise the RFDS, using their radio and emergency signal, but that did not work either.

Finally they were able to get help through a ham radio operator in Indonesia, who had picked up the SOS call from the Marble Bar hospital. He spoke to the Matron, then contacted one of his ham radio mates in Port Hedland who went to the hospital and told the doctors what was happening on *Hillside Station*.

It was with the help of these two men and their interest in radio that, between them, they were able to patch the call through to the RFDS in Port Hedland. David was then able to tell the doctor what had happened. They turned on their regular radio but it was too late. The RFDS came the next morning, but they were too late to save our baby brother.

Father arrives home

Evening had set in and the sun had gone down before we saw the headlights of the jeep in the distance. It was such a relief to us all when Father pulled up at the gates to the homestead. Before our father could alight from the jeep he found himself surrounded by us children all frantically trying to tell him what was happening and begging him to help Mother.

Father and Bob Otway realised very quickly that our mother and her baby were in a lot of trouble and they saw that the birth was imminent as Mother was not going to be able to hold off having her baby for much longer.

After being told that I had not been able to contact the RFDS in Port Hedland as there appeared to be some problem with the transmitter, Father knew that waiting for the doctors to arrive was out of the question; it was now a matter of life or death for mother and child. Menawhile, Bob Otway and Charlie kept trying to get through; using the station's shortwave radio. They must have tried for some time to get help, but the radio emergency signal still appeared not to be working.

Our father must have been just as frightened as we were, but he never let it show as he calmed us children and began preparing for the birth of his child. The nurse thank goodness had started to collect herself, probably because our father was home and had taken over.

While all of this drama was taking place, Mother – who was barely conscious – managed somehow to keep us focused on what was ahead. With Maureen's help Father set about getting Mother ready for the birth.

After checking that the little ones were fed and in bed, Mary and I were told by my father to get out of the way; better still, to go to bed. Maureen was still Mother's greatest comfort, and Mother began to rely on her more and more as the night wore on.

On the night of the 11th May 1954 at about ten or eleven o'clock; little John Patrick was born; he was able to breath on his own for a short time, but then he started having difficulties with his breathing and stopped. Father tried to resuscitate his baby son, and did manage to start him breathing again, but it was only for a short while; sadly John Patrick then slipped away from us forever. Our little one was with us for such a short time.

I had been in bed, waiting to hear what was happening, for what seemed like hours and not daring to go and ask father or the nurse how Mother was, just in case I would be told yet again I was in the way.

I thought father was angry with me for not being able to get through to the doctors in Port Hedland, and the nurse had made a point of telling me a number of times during the ensuing drama that I was useless and it would be my fault if my mother and her baby died.

It was as if a great weight had descended on my shoulders; and I was left feeling totally useless and unable to help anyone.

I found myself creeping along the veranda and standing in the dark just outside Mother's bedroom a number of times as the night wore on. I just stood there in the shadows, wanting to see if I could help, but I was careful to make sure that I was not seen. I had become so terrified of upsetting Mother and Father; and did not want to make things any worse than they already were. I spent the time wondering if this night would ever end.

During the afternoon and evening the nurse had kept yelling at me "Where Is Your God when we need him". I think she had found me praying to God to help us. Very annoying.

It must have been some time after John Patrick passed away that father came down to where we were sleeping. I heard his footsteps so pretended to be asleep. Father lightly tapped me on the shoulder and when he became aware that I was awake he sat on the edge of the bed and, with tears running down his face, told me that he and Mother had had a son and us children a brother. When I asked to go and see John Patrick our father said no, I could see him in the morning. This puzzled me somewhat. Father's tears fell even faster and he told me between sobs that our brother had gone to God.

Our father then left to go back to the bedroom. I was still not sure if our mother had died or not and was too scared to ask him just in case she had died as well. Father was walking around in quite a state and had forgotten to tell me about our mother. I spent the rest of the night praying that she was still alive.

After making our mother as comfortable as they could, Maureen and the nurse tried to rest for a few hours while our parents spent the night comforting each other and cradling the body of their baby son. They also had the distressing task of making preparations for John Patrick's funeral the next day.

My father and Bob Otway and Charlie stayed up with our
father all night and Maureen was kept under the nurse's eye. In
the morning Maureen and I were made to cook, though there
was not all that much to cook anyway. But Bob Otway, it
seemed from my memories that he was a very strong man, and
he took care of the needs of everyone, I think. Bob made the
casket and dug the grave for my mother and father. They were
so broken up. Bob took over, I think.

Early the next morning, just as a weak sun was showing its face
over the nearby ranges, everyone was up and the station had
become a hive of activity with our father and the men preparing
the small station airstrip for the arrival of the RFDS plane.
Father had been able to use the radio at last and the plane
would be bringing the doctor as well as the Catholic Priest
(Father J.F. O'Sullivan) and the local policeman, from Marble
Bar. This was Ian Blair who was a family friend and our
parents had known his family for years. They, like our parents,
were pastoralists up north.

Bob Otway made a little casket using a timber box that he had
found in the store room and lining it with material that mother
had in her sewing basket. Mother placed a set of Rosary Beads
and a flower next to the body of John Patrick. What I can
remember about that day was that John Patrick looked like a
little old man. What a strange memory.

Elizabeth also managed to have a peep at our little brother's
body, which was being kept in the wardrobe in our father's
bedroom, much to the horror of the adults and for her troubles
was soundly told off. Liz was all of five at the time, as she had
just had her fifth birthday in March of that year.

John Patrick's Funeral

The day of John Patrick's funeral was overcast with just a hint
of rain to come.

During the night we could hear the thunder and see the lightening; this set the mood perfectly for John Patrick's funeral.

With the threat of a possible storm, Father and Bob Otway quickly set about organizing the funeral for John Patrick. The big concern was that a storm might do damage to the airstrip which was the last thing Father and his men wanted.

If a storm did hit the station before the pilot and his passengers had left, the rain could leave the airstrip unusable for at least a couple of days after, thereby stranding on the station everyone who had come to help till the airstrip had dried out.

The nurse had organised us older children to bake cakes and make sandwiches for the visitors that day. I can remember one of the part-Aboriginal ladies coming and helping us. I wish I could remember her name. A lovely person who made our mother's life a little easier over the years. My sister Maureen was outstanding in the way she helped take care of our mother. Maureen was wonderful and even now I stand in awe of her strength that day.

It was mid-morning before the Flying Doctor's plane arrived so, after the doctor and the priest had spent time with Father and Mother, we gathered near the rockery garden where a small grave had been dug earlier in the day for John Patrick's little coffin. Bob Otway and Charlie had offered to dig our brother's grave as our father found the very thought of doing that task more than he could face.

Our parents selected the rockery garden as a burial site because it was a favourite spot of our Mother's and where she would spend hours tending to her rockery garden. It was also close to Mother and Father's bedroom.

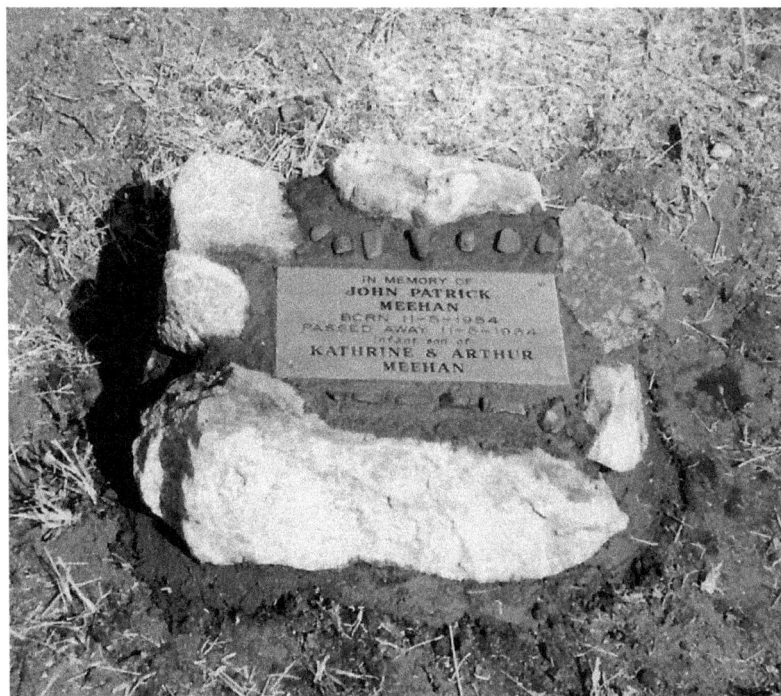

Figure 29: Gravestone of baby John Patrick Meehan at Hillside Station. Mother's name should be spelled as "Catherine".

Our father was supported by Bob Otway and Charlie as well as Ian Blair and Johnno Johnson and his wife Billie. Maureen and the rest of us children stood together, near our father. Mother was too weak and ill to leave her bed but, with the doctor and his nurse by her side, was able to see and hear the funeral service taking place in the garden next to the rockery. Mother's health continued to deteriorate and the doctor was becoming very concerned about her, so plans were put into place to take mother to St John of God hospital in Subiaco, Perth.

Doctor W. Fetwadjieff M.D. was the doctor from the RFDS in Port Hedland who came out to the station to help Mother. With him came the RFDS nurse. These nurses travelled with all the doctors called out to attend to medical problems in the North West.

Father J.F. O'Sullivan was the Catholic Priest who officiated at John Patrick's funeral.

The nurse who was looking after mother before she had the baby was one of the witnesses and Bob Otway (a friend of the family) was the other witness to the burial.

To hold John Patrick's funeral, our parents had to apply for special permission. This was handled by Ian Blair, I think. There was only about twelve hours between John Patrick's birth and his funeral to do this, but with the help of Ian Blair (a policeman at Marble Bar), and Father O'Sullivan as well as Doctor Fetwadjieff M.D.[55] this was done.

Maureen, Judy and Elizabeth were crying during the service and our father seemed lost to us, his children. Our mother was totally unreachable as she had quietly withdrawn from everything around her. It was as if she was with her little boy saying her final goodbye.

Mother's nurse was standing behind my sisters and I, keeping an eye on everyone and everything including me, as earlier that day she had taken me aside after finding me crying behind the bathroom. The nurse gave me a good telling off and said

---

[55] Doctor Fetwadjieff was a doctor from Europe who came to Australia after World War II. My memory of him is not good but I do know he was a brilliant man who was much respected by the people living up North. From what my sister Maureen has told me, like all other doctors from overseas, he had to study and pass medical exams set by the Australian Medical Board before he was able to practice medicine here in Australia. So these doctors had to find jobs where they could both study medicine as well as work to support themselves and their families. If they passed the medical exams they could be sent to country hospitals to work for some years I understand. Apparently Doctor Fetwadjieff worked on the railways to support himself and his family when he first came to Australia.

"Virginia, stop this nonsense. You do not have the right to cry. Your mother and father lost their baby, not you. So don't you dare go crying during the service. By now you must know that there is no God, because if there were a God he would never have let this happen to your father and mother."

After the nurse's lecture I made my way up to the stock feed room and had a good cry there, where no one would see me. But Bob Otway and Charlie had overheard the nurse yelling at me and came looking for me. They gave me a hug and told me it was not my fault that the radio was not working the way it was supposed to. They both gave me a great deal of comfort that day.

It was just after the funeral had finished and we were serving food and drinks to everyone while the doctor and his nurse were preparing Mother for the plane trip to Perth when she had a heart attack and had to be resuscitated by the doctor. It was at that stage that Maureen and I realised that it was going to be touch and go for our mother. Judy and Elizabeth were too young to understand what was happening, yet they seemed to know how bad things were and stayed close to Maureen and me.

It was Bob Otway who came up with a solution for getting our mother to the plane as she was not able to walk, Bob dismantled one of the dining room doors, covered it in a blanket then slid the door under mother who was then strapped onto the door. Father and Bob then carried the door with mother on it out to father's Jeep. Bob drove out to the plane with Father and Charlie holding on to the door and mother across the back of the Jeep. On reaching the plane, the door was manhandled into the plane where mother was then transferred on to a fixed stretcher.

After the Funeral

Mother's nurse decided to travel down to Perth with our mother on the RFDS plane, and waited while Doctor Fortune treated

Mother for her heart condition. The nurse then stayed in Perth for a few days before returning to *Hillside Station*.

Maureen's friend Mary was sent home to her father a few days later, as soon as it could be arranged.

Our father had asked the nurse to stay with him till he felt able to take care of himself and Elizabeth when we went back to school. While Mother was in hospital in Perth, the nurse stayed on the station for a number of weeks taking care of Elizabeth.

The nurse returned to the station about a week after accompanying mother to Perth, and bought back with her a birthday cake to celebrate my birthday which was on the 10th of June. As well, she had balloons, crackers and lots of other goodies. The cake had little blue Birds of Paradise on the top and the nurse suggested that I send the little birds down to our mother in hospital to cheer her up, which I did.

I am not sure if the nurse came up with the idea of the birthday dinner or whether it was Doctor Dick's wife, Nancy, who was a good friend of Mother's. But the adults must have realised how traumatised we all were after losing John Patrick and then nearly losing our mother so they decided to try and make things a little brighter for us all by bringing some fun into our lives. It was the first and last time I can remember any of us ever having a birthday party.

Back to Loreto

Maureen, Judy and I returned to Loreto Convent soon after the nurse came back to the station. Judy came back to Perth with us I believe; I am not sure but I think Judy had already started school at Loreto Convent that year.

My sisters and I were suffering from the loss of our brother; and the fact that we nearly lost our mother as well during the drama that took place on *Hillside Station* during those school holidays. When we finally returned to Loreto Convent, we were in mourning and quite clearly struggling to come to terms with the tragic loss of our little brother.

Returning to Loreto Convent felt to me like we had entered a sort of time warp. It appeared that everything was the same, yet not the same, because we had changed. In that short space of time, just four weeks, the life that we knew had all but disappeared or been erased and I for one was left feeling as if we had been picked up by a strong wind and dropped in a faraway desert.

How worried we were for our mother as we waited each day to hear if she was going to live or die. This was putting quite a strain on us at school. Mother had by then had two more heart attacks and the doctors had told our father that she was critical and for the family to be prepared for the worst as they did not see any chance of mother recovering; they also noted that our mother seemed to have lost the will to live.

Maureen and I were very apprehensive as we waited for the dreaded telegram from our father that would tell us that Mother had died. From memory, that went on for a while. Mother was declared critical by the doctors and was kept in ICU as they worked to keep her alive. The doctors were so convinced that mother was not going to make it that they advised our father to come down to Perth straight way, but he said he was too busy to leave the station. This tore our mother apart for the rest of her life.

It was touch and go a number of times and each time the hospital would contact the nuns at Loreto and tell them to prepare us because our mother was close to dying and to let us know that the priest had been and given her the last rites of the Catholic Church.

The school principal, Mother Rosalie, made a point of taking Maureen and myself aside after lunch one day – it was not long after we had returned to school – and told us that we were not to discuss anything that had happened on the station during the school holidays or tell anyone that our brother John Patrick had died; it was a taboo subject as far as the school was concerned. No sympathy or understanding was ever shown.

Maureen and I both protested but it was all in vain as we were dismissed after being told to go back to our classes. As an afterthought – or so it seemed to me at the time – Mother Rosalie added that the hospital had rung to say that our mother's condition had deteriorated and that we should come to the hospital straight away. The doctors did not expect Mother to last the day.

Maureen and I both begged Mother Rosalie to let us go to the hospital so that we could be with our mother, before it was too late. We both stressed that we wanted to be with Mother, as we did not want her to be alone if she were to pass away.

Mother Rosalie told us very firmly that we would not be going to the hospital to see our mother, as she (Mother Rosalie) thought that it would be inappropriate. Then we were told to go to the chapel and pray, but not before being reminded that we were not to mention the matter again.

At last our mother started to fight back and after many weeks in hospital went on to make a rather tenuous recovery. However, she was left with a number of medical problems to deal with. Our mother was never the same person again. It was as if a part of her had died with John Patrick.

Talking to my sister Judy when I first started writing this story, and while Judy was still alive, her memories were very much in line with what I have written. Judy always said she felt she was very much affected by the events played out on *Hillside Station* on the 11th of May 1954.

We were all fragile and in pain and I believe in the end we shut down. That was the only means we had left to us to survive.

I believe to this day that Maureen and I were emotionally and psychologically damaged, as were our two younger sisters, Judy and Elizabeth. Even though they were much younger than we were when this all happened, the experience impacted just as strongly on them as it did on Maureen and myself.

The drama that took place on *Hillside Station* that day meant that we have each had to find our own way of dealing with these past memories.

My reaction to the events of 1954 was, on reflection, quite dramatic as it ended up affecting my schooling. I more or less shut down and most times did not even register what the teachers were saying. To cope with school, and the stresses at home, I made myself an emotionally safe place to hide. There were many times when as a child I would shut the world out. I believe now that I may have suffered from Post-Traumatic Stress Disorder. My sister Judy also felt that she suffered from the same thing.

Grieving for John Patrick

Our family suffered deeply from what had occurred on *Hillside Station*. Father and Mother discouraged us from talking about John Patrick and his death. It was not put into words – just implied – and when I did try on the odd occasion to mention it, I was silenced very quickly with a look that said much more than words. It was just too painful for them both, I believe.

Sentences were left unfinished and words just seemed too hang in the air when around our parents. This was to have a profound and lasting effect on the whole family.

Our parents' relationship started to form serious cracks as they went their separate ways emotionally. They seemed to be unable to support each other after Mother came home from the hospital.

Father had by that time found a way to handle his grief and move on. Mother, on the other hand, had gone through her own hell after John Patrick's death and had not really been able to mourn him properly.

I believe Father did not understand this and thought our mother needed to move on like he had. He could not – or would not – understand that he had replaced the pain of losing his son John Patrick by turning all his energies to working on the station. This was his way of coping and also a way of easing the stress that he was under, as by night fall he was so tired he would fall asleep straight after his evening meal. That way he did not have to talk to Mother, or to us children if we were home.

Mother on the other hand spent hours just standing next to the grave and weeping for what she had lost.

When home from school it was quite common for me to find Mother standing alone by John Patrick's grave quietly weeping. When I asked if I could help in any way she would just look straight through me, then turn and walk way. I soon learnt to just leave her to grieve alone.

I felt then, and I still do now, that both father and mother blamed me for their son's death. If I had got them help he may have lived, but we will never know. Both my parents developed a blind spot when it came to me and in many ways I became invisible to them both. When it was possible they did their best to pretend I was not there.

As an adult, I understand that what had happened on *Hillside Station* was not of my making. I had – as an eleven year old child – done what I could to help my mother, but it is only now, as an adult, that I have come to this understanding. As a child I was tormented by the belief that it was in some way all my fault, and my parents never at any stage attempted to address that misconception. Because of our parents' inability to cope, my sisters and I found that our relationships were also damaged.

I believe mother needed to blame someone for what took place on *Hillside Station*, even though she knew that what she felt was unrealistic. Most of all she blamed herself and carried that guilt to her grave. Mother also held our father responsible for her loss, because he had demanded that she return to *Hillside Station* and wait there till she was nearer the birth of her child. She was also somewhat resentful that he was not at home on that fateful day. Unfortunately I believe our father also held mother responsible for the loss of his son and heir.

To my knowledge, neither of them talked about what had happened, so were probably unaware of what they were doing to each other and us children. Back then there was no one that they could turn to for help and comfort. I feel at times that they both must have felt overwhelmed by their sense of isolation and loneliness, as they had no family or friends close by who would have understood what they were going through.

Father and Mother suffered for the rest of their lives; it appeared that they had come to believe that they were each on their own.

Unfortunately, because of the small population that lived in the North West at this time, there was also a severe shortage of medical and support services for people in need. They had to travel to Perth for any extra medical help although there were some wonderful doctors who came north to Port Hedland. Many were immigrants who came from Europe after World War II. The Australian Government of the day would send these doctors to country areas for a time.

Because of the lack of professional services available to those who lived in the North West during this time it was very difficult for those people who had developed health problems, and also for pregnant women to get medical support and to have tests carried out if they were needed.

The women who made this choice to marry and live up North were very strong and independent ladies who mostly adjusted well to station life on the whole, but for our mother it was not to be.

Police Brutality

After the death of John Patrick, our parents were faced with an even more callous and traumatic experience when they received an unannounced visit from a Port Hedland detective. This very ungallant policeman came alone. He had been sent to help Mother and Father fill out the required paper work regarding John Patrick's birth and death.

From what I can remember Father and Mother telling us, it was a very harrowing experience for them both.

Father had filled in the information about Father and Mother's children to Ian Blair. I have copied father's list below.

> Mary Philomena deceased 1938
>
> Maureen Patricia 1941 12 years
>
> Margaret Frances Virginia 1942 10 years
>
> Judith Anne 1947 7years.
>
> Elizabeth Carmen 1949. 5 years
>
> John Patrick Baby. 11th May 1954. Just born.

Unfortunately; the policeman that was sent out to *Hillside Station* on that day treated his visit to our parents as if his mission in life was to solve a murder case and Mother in particular was made to feel as though she had indeed committed a crime. He was callous, crude and totally without compassion towards our parents.

The questioning went on for hours as he asked mother not only about John Patrick's death but also about Mary Philomena's death in 1938. He made it very clear to mother that he was treating John Patrick's death as if it was a homicide – which of course it was not – and if that policeman had bothered to talk to Ian Blair in Marble Bar he would have been made aware of the situation and shown all the documents.

Mother had not long come home from hospital when this happened. This unwelcome visitor left Mother shattered. She was totally destroyed and never recovered from that policeman's visit. He had made it very clear that he felt Mother had in some way contributed to the deaths of both her babies.

Father complained to the police in Port Hedland about what this detective had done to them both and demanded that the police force do something about him. The police there were very apologetic to both our parents.

Father and Mother found out much later that this particular policeman had in fact been sent to Port Hedland for disciplinary reasons; I believe that it was not long after the visit to *Hillside* that he was moved again.

It was from that time that Mother and Father both began to suffer from a depressive illness from which neither recovered.

Father and mother were remarkable people and in their own ways had achieved amazing things during their life together. They were well respected and admired for their courage and strength, so it was very sad to see them suffer so much after John Patrick's death. They deserved more, they deserved better.

.oOo.

CHAPTER 15
Character Building

Our father and mother were very much into what they thought of as building up our characters. This could take many forms, from smacking to verbal discipline, but the main recipients of their character building were Maureen and I.

There are eleven months between Maureen and myself and looking back, I think this must have been very hard for our mother because she would often find herself alone with us two babies who were both under two years of age, at the time. This was in the early 1940s.

The family lived in the Murchison (a pastoral area of Western Australia) which to our mother must have felt like the back of beyond. To make matters worse for our mother, there were no family members close by to give her support or help with two young children. It must have been a very harrowing experience for our mother and this unfortunately led to lack of bonding with her two young children as well.

Our father had his favourite children who were Maureen and Judy. Mother also had her favourite child, and this again was our sister Judy. Mother had a close bond with Judy and to a lesser degree our youngest sister Elizabeth. As children – and even when we were adults – we were always encouraged to be competitive with each other, but not in a healthy way.

Father and Mother believed that bullying was a normal part of growing up, and they viewed it as part of our development. To an outsider it must have appeared that our parents encouraged bullying between their children. I can remember that our father saw it as a means of one child asserting her dominance over another, and therefore a good thing.

It was very much frowned upon for the target of this bullying to go to either parent for support. They considered it a sign of weakness by the child.

Judy was the only one from memory whom our parents protected from being bullied. She was considered to be very special to our parents and not in need of their character building treatment.

Elizabeth, on the other hand, was not so fortunate. A beautiful and engaging child, Elizabeth still came in for her share of beltings from our father and mother while growing up. Usually administered by our father, I am sorry to say.

Our parents believed very much in the strap, which our father used with great force till we reached ten years of age and then the belting would stop as our father believed we had come of an age to be considered too old for beltings from him. Our mother, on the other hand, believed in using whatever she could get her hands on at the time.

Mother discouraged us from forming close ties outside the family unit, especially after she lost her last child, John Patrick. Mother's sisters – and especially Aunty Nell – tried very hard to break through Mother's self-imposed solitude. Nell, bless her, never gave up trying to help Mother.

Our parents also saw value in their children competing against each other. This made sense at the time, as there were no children of our age close enough for us to play with and that was why our parents decided to send us away to boarding school in Perth.

Our neighbours that had children our age lived many miles away on other pastoral properties and the distance between us was too far for regular visits. So we rarely saw other children apart from Tadgee and Jerry until we went away to school in Perth.

Self-defence

Maureen and I had a rather tempestuous relationship while we were growing up and, as I have said, we were always encouraged to compete against each other for our parents' attention. Fights were common between us and could became quite heated at times

One day, our parents had a visitor to the station. I think his name was Bob; not sure but I think that was it.

Bob was the new Government dogger for the Murchison area. He was employed by the Agricultural Department of the Western Australian Government to reduce the number of dingoes and feral cats and dogs on the pastoral properties; this was done by both baits and traps.

The baits were made from the carcases of dead kangaroos or a dead calf or lamb. The dogger, would cut up the dead animal to use as bait. He would place a small amount of strychnine inside the meat, and this would then be attached to the trap; the trap would then be placed near a watering hole in the hope that the dingo would be attracted to the smell and would then take a bite and so end its life. Not a nice way to die for dog or man.

The traps were made from a heavy metal and were about a metre long. One end had a long chain that could be pegged and buried in the ground. At the other end there was a row of very sharp teeth. These would clamp shut on any animal that tried to take the bait, crushing the animal's leg or head when it tried to take the meat. The trap was buried in the ground with just a small amount of meat showing to entice the animal to try. The dingos could and would put up quite a fight to get free from the trap and many times they were successful

The dogger would also use the whole body of a dead animal; he would salt the body with a good amount of the strychnine and bury it in the ground with just a leg or head showing to attract the dingo. In later years, the men would also use a poison called 1080 pellets which were deadly but worked very well.

I can remember our father (who also did his own baiting and trapping) kept the traps and poisons in a special room which was always locked. This was next to his work shop and it was a no-go zone for all but our father.

Bob stayed at *Mt Vernon* for a number of days and would go out around the property with our father as he learnt where the trouble spots were. That is, where the dingoes were causing the most damage to Father's stock. The dingoes would attack new born calves, a day or so after they were born, or lambs which were another favourite food source for the dingoes.

A few days after arriving at *Mt Vernon*, Bob witnessed, a fight between Maureen and myself which he found very disturbing. After quietly observing Maureen and my relationship, and taking in the fact that our parents appeared to be encouraging it, Bob spoke up, telling them that this was not the way children should behave towards each other and that they should intervene and instead encourage us to became friends.

Bob became enemy number one with our parents as they were not impressed that he should take it upon himself to speak out. They both considered it in bad taste that he had spoken up and they let Bob know, in no uncertain terms, that it was none of his business.

One of Father's and Mother's most used comments in regard to Maureen and me was "Well, Virginia has to learn, sooner or later, to stand up for herself". Bob was not that impressed with their thinking, so set out to teach me to defend myself. After telling both our parents what he planned to do, and expecting their support and interest, Bob was somewhat surprised, when our parents showed no interest. Bob took me aside and tried to teach this four or five year old self-defence. From what I can remember, it worked up to a point.

The next time Maureen started to push me around, I picked up the nearest thing to hand, which was an old rusty tin can, and fair let her have it across her poor head.

Well, there was blood running down Maureen's face from her cut head, and for us young children it was a terrifying sight. Blood seemed to be everywhere. Maureen started yelling her head off. The Aboriginal ladies came running and Maureen took off for our mother. Father heard the ruckus and came running to see what had happened.

He saw Maureen with a bloody head and face and turned to me, who had guilt written all over my face; but still following close on Maureen's heels. I was not going to miss out on this great big drama even if I had caused it.

That is, till I saw our father's thundering face, and heard his and our mother's angry yells. Drama or not; I knew I had to find a safe place to hide till every one calmed down, or I would be caught and given the belting of my life.

Father picked up his strap and I took off, with Father close behind. Luckily for me, Father had not stopped to put his boots on. Even luckier for me, he ran into some prickly plants. This slowed him up somewhat. In fact he stopped and decided to return to the homestead to get his boots. Then he planned to come after me again. This gave me the time I needed to reach the safety of the Aboriginals' living quarters where Fanny and the other ladies hid me until Mother and Father had calmed down. Or, more to the point, till the ladies thought it was safe to let me go home. They were not going to let Maureen or myself be punished for what had happened.

Meanwhile mother used her nursing skills to patch Maureen's head and to calm Father down and set his mind at rest that his eldest daughter was not going to die any time soon. Maureen still has the scar on her head from the rusty old can that I hit her with.

Needless to say I was not allowed any more self-defence lessons.

Getting Back on the Horse

One time when our father was away mustering cattle, and there was only our mother and us children at the homestead, we had a friend of Father's visit. He was from a neighbouring station, and he arrived on horseback and leading a pack horse. I cannot remember the man's name but he was a regular visitor to *Mt Vernon* during our time there.

My memories of this man are not all that good but we must have liked him a lot, because I went running up to the saddle room where he was unsaddling his horses to greet him and receive a big hug.

After a while he picked me up and put me on the back of his stallion with the intent of walking the horse around with me on its back, which he would do on other visits to the station. This all took place just outside the saddle room.

Just down from the saddle room and the other work rooms was a group of very large boulders. This is where the story gets a little bit more interesting.

Unbeknown to this friend – and of course to me – the station dogs had gathered to watch the show and to see what they could do to have some fun. As soon as I was settled into the saddle, two of the dogs ran in and nipped the stallion on his back heels. With that, the horse took off, pulling the reins out of the hands of the visitor and leaving me, as a five or six year old, hanging on for all I was worth, and our friend running after me and the horse trying to catch him and me before I was tossed off.

After a rather hectic ride, I was finally tossed by the horse. But unfortunately for me it was right into the boulders. I was lying on the ground stunned and crying. Our mother came running to see if all was well and then said "Well if your father was home he would make you get back on the horse straight away". This did not go down all that well with their friend, but Mother, being Mother, insisted, and back on the horse I went. Again, no one had thought to take care of the dogs; so in they came again and nipped the horse on his heels and once again, the horse took off, with me clinging on as hard as I could till finally I was tossed off yet again and it was straight back into the same boulders.

Mother and her friend looked at each other and the friend said "That is enough, there is no way I will put her back on that horse again." With that I was told to go inside and wash up. Fanny took care of my cuts and bruises and wiped away my tears, and I can still remember the boulders that lay around the homestead. I don't think Mother ever told our father about that misadventure.

Smacking Sisters

One of the more harrowing character building experiences that I can still remember with guilt and dread was when our father decided that I had to discipline my two little sisters, Judy and Elizabeth.

To this day I have no idea what these little girls had done to upset our father and I doubt that they had done anything wrong at all.

I was about twelve or thirteen at the time when one day, after Father, Maureen and I had come back to the homestead from working around the property, Father took me aside and ordered me to spank my two sisters. I was horrified and protested but it was in vain because when our parents told you to do something, you did it.

Unfortunately and unbeknown to me, my sisters had earlier been told by our father what he had planned for me, so they were instructed to put a book down their pants so that when I hit them it would hurt me more than it would hurt them. Take my word it did, but the humiliation was even worse, thanks to our father. The family were all able to have a good laugh at my expense.

There is a sad ending to this little tale. Elizabeth and Judy never forgot or put to rest what had happened when we were young and I do not for one moment blame them. They had their faith in me – their sister – shattered for good, thanks to our father.

Prayers

Elizabeth was also on the receiving end of another of Father's character building ideas. Poor Elizabeth was quite young at the time and had not yet been sent away to Loreto Convent. She was possibly about five or six years old when she was taken to task by our father for – of all things – not saying her prayers. Now my parents could hardly be called religious people at any stage in their lives, so father taking Elizabeth to task for not praying could only be called strange even by our parents' standards.

On the day that this took place, father had come home from working around the traps (that is, working around the property). He was upset and angry about something that had taken place while he was working. Possibly problems with windmills or downed fences, possibly even dead cattle, who knows?

The family never did find out what had upset him so much but whatever it was, Father was on the war path, looking for someone or something to take his temper out on. Elizabeth was the first person that father saw when he got home all fired up so she felt the full force of her father's strap on her little bottom. Elizabeth deserved better from our parents – especially her father – than that.

.oOo.

CHAPTER 16
Life Skills

Like our grandfather, J.P. Meehan, our father took it upon himself to teach Maureen and me some life skills. However, he was not cut out to teach children; he just did not have the patience. Learning to swim, or ride horses, or even to handle guns came easily to our father apparently. Also, when he was growing up, his father – J.P. Meehan – must have been much more patient when he was teaching his sons and daughter these skills than our father ever was with us.

But being taught anything by our father – as a young child, or even a teenager for that matter – could be terrifying, and I remember always trying to hide when father felt the urge to teach. That stayed with me all my life. Mother finally stepped in and put an end to Father teaching us riding or swimming. I am not sure if Maureen learnt to ride later or not, but I certainly did not.

Horse Riding Lessons

First it was riding. I have mentioned this before, as a friend of the family had given us a little Shetland pony which our father thought was a good starting point for giving us riding lessons.

The first lesson that we learnt was to head for the hills as fast as we could, and wait there till Father found something else to keep him amused.

Horse riding was one of the skills that our father did not have the patience to teach; swimming was another. Which was a bit surprising as father was a strong swimmer himself, as well as being a very talented horseman. Our father thought all he had to do was toss us into the river, or sit us on a horse – any horse – and we would become proficient at whatever he thought we should be able to do.

To our father it was "Jump on the horse, stay on the horse, or else.". The 'or else' meant I for one was forever falling off, partly caused by our father's constant yelling at me. Yelling was just part and parcel of our father's nature. Which was OK for him, maybe.

Tadgee, like us, was also 'taught' by our father to ride and swim but he did not have the patience with her either so mother and Fanny put a stop to that as well.

Jerry was lucky as he was taught by the Aboriginal men to both swim and to ride horses and other traditional activities that the men took part in.

Swimming Lessons

One time, our father felt the urge to teach Maureen and me how to swim. There was a river (the Ashburton I think it was called) about two or three miles from the *Mt Vernon* homestead where we would go for picnics and a swim. Mother would pack a picnic lunch and Father would take us all there in his ute. It was always an enjoyable time for the family.

This day father decided that it was as good a time as any to teach us both to swim so he tossed a horse hair mattress into the back of the ute and away we went. Mother agreed in principle with the swimming lessons, except that she had not anticipated how father had planned to teach us. Father was a good swimmer and in fact had saved someone from drowning in the Swan River in Perth when he was a young man. For us, the swimming lessons were heading towards another non-event for me and Maureen.

After lunch and a bit of a paddle, Father set about teaching Maureen and myself the fine art of swimming – that is, by his method of teaching – which turned out to be quite different from most folks.

Our sister Judy was just a baby at the time, so our mother found a cool spot for Judy's bassinet and settled in for a relaxed afternoon. Mother had brought a book with her so sat back to read and relax.

Figure 30: The Ashburton River at Mt Vernon in a dry season

Maureen and I thought being pushed around on top of the horse hair mattress was great fun and we kept on asking our father to keep on pushing us. Father pushed the mattress with us still on top into the deeper part of the river, and we thought this was great too. Until, without warning, Father flipped the mattress over. Maureen and I went flying off the mattress and into deep water.

Maureen and I had never had a swimming lesson, so had no idea what to do. We sank like two little rocks, yelling our heads off for Mother. Needless to say, our mother was not impressed at all with our father's teaching methods, yet again. Maureen and I would have been about six and five years of age when this happened.

We never had another swimming lesson – or, for that matter another horse riding lesson – from our father. For him, that was the end of his interest in teaching his two little girls to swim or horse ride. Mother told him his methods were very questionable, so our father stormed off in a huff and that was that for us when it came to him teaching anything to do with station life or life in general.

Maureen spent some time on *Hillside Station* after she left Loreto and again after she left nursing school at the Royal Perth Hospital. This would have been before she and Don Stubbs were married; maybe 1961. I understand that our father was very grateful to have Maureen with him during this time, so he most likely taught her a lot about station life and how to manage the cattle that he had, because Maureen now has her own farm at Bindoon and does a wonderful job.

Shooting Lessons

Father even taught us how to fire guns – 303s and 22s – not sure if that was what they were called. Our father also had a couple of revolvers including one that he used to take around with him on the property. I can remember once being given the use of the revolver to shoot at a target that our father had rigged up on a tree for us to use for practice. Maureen was very good and still is today. Me; well that was another story altogether. I hated hurting or killing anything, so was a dead loss in Father's eyes. A bit too soft, he thought.

I hated shooting at live animals so would miss on purpose, much to my father's annoyance, as he told me I was wasting bullets, which would have been true. One time he had Maureen and me shooting at galahs that were stripping leaves from the house trees, I did not want to do this but our father insisted so I tried to miss which ended in one poor bird having a damaged wing so the family kept it as a house pet for about a year.

Figure 31: Arthur Meehan with his jeep on Hillside Station

Driving Lessons

I can still remember asking my father if he would teach me to drive his Jeep – the only motor vehicle on the station at that time. I was home after leaving school the year before.

I thought my father was going to turn himself inside out. Anyway he told me in no uncertain terms that there was no way on God's earth was I going to get behind the wheel of his Jeep. Years later, when I was living in Perth and Father had come down from up North, he asked me to drive him to some appointments he had. Later he told me I was in fact a good driver.

Teaching bike riding

There were good times as well when Maureen and I decided to pass on our skills, learnt from our father, to both Judy and

Elizabeth. Now, remembering that our skills were only half learnt from our father so it made for interesting times for Judy and Elizabeth.

We had acquired a rusty old bike (from Ian Blair I think it was). It was a pretty clapped out old thing but Maureen and I both learnt how to ride it. Then Judy and Elizabeth were given a bike by our Uncle Jack Meehan who lived on *Austin Downs*. Maureen would talk Judy and Elizabeth into letting us use their bike and, when we had the two bikes, we would go for long rides into the bush.

Judy and Elizabeth were not all that interested in the bike to start off with, so Maureen and I made good use of their bike, promising to teach them how to ride it if they let us ride their bike too. Judy was not all that keen to learn but Elizabeth was, so Maureen and I decided that teaching them how to ride a bike was a place to start.

Then came the time when our promise to our sisters had to be fulfilled. We started with Judy and that went quite well. She stayed on and remembered to use the brake when she wanted to stop. Then it was Elizabeth's turn.

Elizabeth was somewhat younger than Judy, but Maureen was quite confident that we could teach her, so away we went. First we walked her around the front yard holding the bike by the seat. Then, as planned, we held on to the bike to keep it balanced so that Elizabeth could ride the bike with just a little help from us. When we felt that she had become more confident, we planned to let go of the bike so that our sister could ride the short distance to the house on her own.

By this time Elizabeth was showing signs of taking it all in, so Maureen thought now was as good time as any to let her ride on her own. So we let go of the bike and stood back to watch Elizabeth continue on her own, from the front gate to the house.

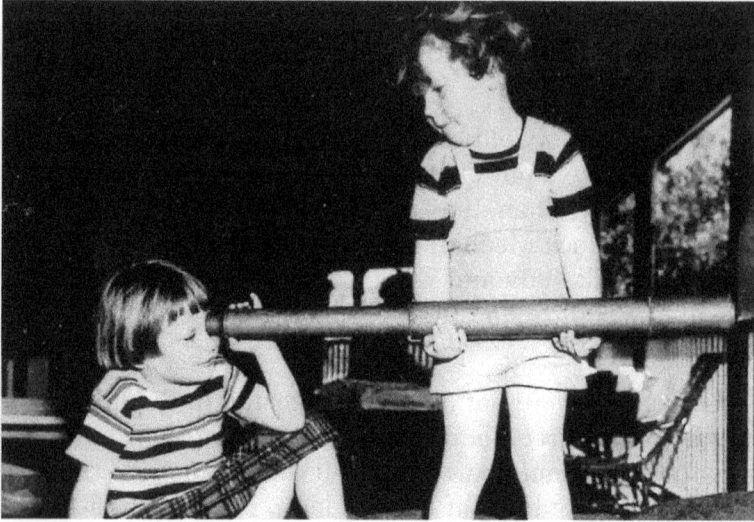

Figure 32:Judy Meehan (seated) and Elizabeth Meehan

We forgot to tell her what we had planned, so Elizabeth was totally unprepared when she set off and away she went. She was fine for a little while but then she realised that we were not holding on she started to panic and was yelling "I am going to fall; I am going to fall."

Maureen and I took off and were running as fast as we could while yelling at Elizabeth to put on the brake, but she had forgotten all that we had shown her that morning and was heading straight for the Poinciana and Jacaranda trees that lined the garden path leading to the house. Maureen and I just made it to Elizabeth before she crashed into the trees. We thought it was very funny after we had caught our breath but Elizabeth was not impressed with us as teachers and she would decline our offers to teach her in a very robust manner. It was a while before she would allow us to get her back on the bike.

.oOo.

CHAPTER 17
Droughts and Cyclones

Droughts and cyclones were common. While our family lived on *Mt Vernon* and *Hillside* stations we experienced cyclones most years and droughts came frequently.

Droughts

Some droughts lasted for years. Our father told Maureen and I that we were both born during a long drought period.

I remember hearing our parents talking about how bad the current drought at the time was, and our father would refer back to the droughts during the 1930s through to the 1950s when the family lived on *Mt Vernon*.

Mt Vernon had not had any decent rain for years and things were becoming critical, both for the cattle that Father was breeding and for the native animals that inhabited the land. It was not unusual to have a kangaroo or emu coming into the homestead and grazing on the grass there.

Because of the lack of water, the cattle and horses were getting trapped on the muddy river banks when they went for a drink of water. The animals ended up dying unless Father or one of his men found them in time.

Where the animals were often trapped was on the banks of the river where the mud had become very slippery from the constant movement of animals looking for water to drink. Sadly, once the animals became caught they became frightened and would start to struggle; that was when the mud could become their enemy. The mud sucked the animal's feet further and further into its muddy depth. From there the trapped animals would be unable to escape without the help of our father or one of his men.

Those that did not die in the mud were saved by Father and his men who would spent many hours trying to free both cattle and horses as well as native animals that they found trapped.

Father and his men did their best to save the trapped animals but unfortunately not all animals could be saved and those that they could not save had to be shot. Father found this very upsetting and I can remember him coming home very upset many times.

There was little feed left for the animals to eat as the native grasses and spinifex were dying off. The rivers and water holes were drying up, and so were the wells that Father had put in.

While the drought dragged on our father had to keep moving the cattle and horses in the faint hope of finding feed for them all. If the stock were away from the watering holes of the nearest tanks and troughs then they had to rely on the rivers and creeks for their water.

Both the homesteads that the family lived in (*Mt Vernon* during 1930-1951 and *Hillside Station* from 1952-1963) were affected when there was a drought because it meant that the homestead could face a possible water shortage. So all us children were taught to be very careful with what water we had from an early age. Even Mother's garden had to be replanned and she used water from the bathrooms and the laundry to keep her vegetable garden going.

Luckily *Mt Vernon* and *Hillside* stations were both on an artesian basin, so the wells never really dried out. They just dropped lower and lower and Father would have to go and dig out the sand base of the wells which, from what he said, was a very difficult job at the best of times.

Maureen can remember (while we were on *Mt Vernon Station)* trying to save the little finches that were falling out of the sky because of the heat. Birds were starving and unable to find water. Most times the birds died because of the heat.

Maureen asked our father if he could help us, so he built a type of cage where Maureen and I tried our best to help the birds. But as our father kept on telling us, we would have great difficulty trying to save them in the summer heat. Seeing birds falling out of the sky was a sad but common sight during the summer months when the temperature would go well above 40°C and to 52°C plus, in the shade.

Cyclones

The droughts would finally be broken by a cyclone; usually a big one. Maureen and I would have been about two and three years old when *Mt Vernon* was hit by a large cyclone. I think that this may have been in 1944.

One of my earliest memories would have been when I was about three years old. It was the monsoon season up north, so cyclones were a common event during the summer months.

This day was to start out no different from any other summer day. There were a few dark clouds overhead and the wind was starting to blow a gale but that was fairly normal for that time of the year.

It was about mid-afternoon when Father noticed that the wind had changed and was starting to build up quickly. By this time the dark clouds had increased and were now joined by crashing thunder and bright flashing cracks of lightning.

Father turned to our mother and declared that we were facing a cyclone so it was time for him and Mother to batten down and take cover as they waited out what was to come. Luckily for *Mt Vernon* this cyclone which passed over the station homestead did not do a great deal of damage, but it left damage across the station property. There were fences and shedding blown down or away and some stock losses but in general we were lucky. Not so other properties where the cyclone had done more damage.

This of course did not bother Maureen or myself, as we were enjoying ourselves and being so young were totally unaware of the dangers we were facing in the cyclone.

Maureen and I had never seen rain like that before, as the station had been going through a very long drought which lasted for many years. Our only experience with storms was when the station had what was known as a dry storm. This is an electrical storm with lots of thunder and lightning but little if any rain comes with it.

Maureen and I were of course truly fascinated by the events taking place in our back yard. Rain like we had never seen the likes of before came pelting down and both Maureen and I thought this is truly the best fun we had ever had, running in and out of the storm.

Our mother was not so impressed with all the thunder and lightning, though she welcomed the rain very much. But our poor mother had a real fear of the electrical storms, so she disappeared to hide under her bed till the storm had passed.

Maureen and I were running around and yelling our heads off, and wanting to know what this wet stuff was that we were playing in. I think in a way we were quite terrified by the noise from the thunder and the cracking and flashing of the lightning coming out of the sky, but the rain was winning out and we refused to go inside for either Father or Mother. I can still remember our father picking us both up – one under each arm – and making a run for the cover of the house, all the while telling us off. Father and Mother would often recite that story to my sisters, especially if there was a storm raging over head.

As welcome as the rain was, it also created its own problems because of the strong and devastating winds that came with it.

Hillside Station had its fair share of droughts, and cyclones as well, while we were living on the station. Because of the strong winds that came with the cyclones, there was the possibility of damage being done around the property, with windmills and fences being the main areas which were damaged. From memory, the homestead was not damaged by cyclones during our time there.

We had to make sure that everything that could be blown away was locked down and made secure. When we knew a storm was heading our way it was all hands on board to help with the lock up. That way it would help minimise any damage from the cyclones. This could at times be a very scary job. Maureen and I were in charge of making sure everything was ready for when the storm hit if our father was away from the station homestead.

Maureen also got the job of dragging out any stock that had got caught in the riverbed before the river started to run from the rain.

When we had strapped everything down that we could, the family (minus our mother who was still under her bed) would gather in the kitchen with the wood stove keeping us warm and the kettle on as our father would not let us turn the generator on while the storm raged around us. This could be both exciting and scary listening to the wind which sounded like a massive train flying over the top of us.

.oOo.

CHAPTER 18
Challenging Times

Our father began to withdraw more and more after he lost his baby son in 1954. The station was to become his lifeline and he found himself directing all his energies into keeping that operating. *Hillside Station* at that time was experiencing a number of other problems as well.

Pastoralists in the Pilbara region were being affected by droughts during the 1950s and 1960s. *Hillside Station* suffered as did most other station owners in the North West at that time

Figure 33: Hillside Station country (Photo Bree Krieger)

Father had also discovered after signing the contract and taking over the management of *Hillside Station* that the inventory for the number of sheep he had been sold was incorrect or – more to the point – an outright lie. There were about half that number of sheep when our father held his first muster after moving in.

This turned into a long legal battle for our parents and at the worst possible time in their lives. The dispute ended up in the courts and our father was awarded compensation, but it all took its toll on the family.

Father then made the decision to change from sheep to cattle. This decision was made in an effort to try and save our parents' livelihood. Our mother was worried about our father going back into breeding cattle and tried talking him into selling *Hillside*

Station. I believe they both gave this idea a great deal of thought over some months, but father came to the conclusion that it would not be a good proposition for them at that time because of the long drought that *Hillside Station* was experiencing as well as the low wool and beef prices that producers were getting for their stock. They were also made aware that the market for selling stations was very sluggish at that time, and they were advised against putting *Hillside Station* on the market.

Atomic bomb testing

Hillside Station had also developed other unexpected problems – unknown to our parents and indeed to other station owners in the Pilbara area at that time – from the possible adverse effect that the atomic testing carried out on the Monte Bello Islands during the 1950s may have been having on the land.

Our father and mother believed that the bombs being tested on the Islands may have caused some form of contamination to their station which is in the Pilbara area; the possibility that this may have occurred needs to be acknowledged. The drift of debris from the dirty bombs may well have contaminated pastoral properties in the North West.

I know my parents were very concerned at the changes that they were seeing with their cattle and sheep (as were a number of other stations owners) at that time. Unfortunately, they were all kept out of the loop and given no answers to their many questions from what I remember. The effect on cattle and sheep properties during the 1950s has – to my knowledge – never been documented. Here is some information I found about the Monte Bello tests.

> Between 1952 and 1956 the British and Australian Governments decided to carry out atomic testing on the Monte Bello Islands. These islands are situated just off the coast near Onslow, a small rural town in the North West.

The first atomic test that took place was a plutonium bomb called "Hurricane". The reported yield was 25 kilotons; this was placed on the Frigate *H.M.S. Plym* which had been moored in shallow water in Main Bay close to Trimouille Island for that purpose and then detonated on the 3rd of October 1952.

Further atomic testing was carried out in 1956, the "Mosaic G1". This was a 15 kiloton plutonium Bomb. This testing took place on the 16th of May 1956, and was carried out off Trimouille Island. The last atomic testing at the Monte Bello Islands took place in 1956. The information I have found said "Test Height and Type: Tower, 31m. Yield: 98kt".

This was the highest yield test ever conducted in Australia. Since the test yield broke an assurance made personally by the British Prime Minister Anthony Eden to the Australian Prime Minister Robert Menzies that the yield would not exceed 2.5 times that of the Hurricane (Thus about 62kt), the truth was quite different; the true yield was concealed from our government and the Australian people till 1984.

There was a further bomb tested a month later on the 19th of June 1956. This was the "Mosaic G2" off Alpha Island, which is also in the Monte Bello Islands. The reported bomb yield was 60 kiloton (but in fact it was a 98 kiloton plutonium bomb).[56]

At this time of the first test father had just sold *Mt Vernon Station* in the Murchison District and he was waiting to take over the management of *Hillside Station* in the Pilbara. Father was in Onslow at the time and was approached by one of the local station owners who asked father if he would help out with their yearly muster. (I think the name of the station owner was either Barrett-Lennard or Forester; they both had cattle properties in the area).

Father bought a caravan and tents and set up a camp for his family. Maureen and I were put into boarding school at Loreto

---

[56] Monte Bello Island Safaris – About – The Atomic Testing, http://www.montebello.com.au/nuclear-testing

Covent in Perth. Ada Broadley stayed with Mother and Father and helped our mother look after our two little sisters, Judy and Elizabeth.

From where father had set up his camp site they could see the large mushroom cloud that formed after the first bomb was detonated on the island. The cloud was blown in towards the mainland and appeared to hover for some time before dispersing and blowing away from the Onslow area. The understanding was that the mushroom cloud would be blown out to sea, but instead there was a wind change which caused the cloud to be blown back towards the mainland.

This also happened when they detonated the next two bombs in 1956. It was after the June testing in 1956 that they discontinued the atomic testing on Monte Bello.

Father and his family helped out with the mustering for a number of weeks from memory. Then they moved to *Hillside Station*.

During the time we lived on *Hillside Station*, further atomic testing was carried out, and, as had happened before when they were at Onslow, we would see big mushroom clouds pass over the station and surrounding areas after each bomb had been detonated.

It was quite eerie to see and made the stock and even our work dogs very edgy. This would only last for a short time and it always affected the light level as the cloud passed overhead.

What no one in the towns or on the stations knew, but found out many years later, was that the atomic bombs that the British and Australian Governments were testing were what would later be described as dirty bombs, meaning that they were high in radioactive materials and other very nasty toxins.

People living in the North West were never told of the dangers associated with the testing and the possible damage it may do to them and their properties. From what I can remember, our parents and their friends said they were never advised about what action they should take to protect themselves and their families.

I do not ever remember my father ever checking buildings on *Hillside Station* homestead after there had been a bomb let off. I can still remember the big black mushroom cloud flying over the homestead station, not knowing just what was going on. It was very spooky watching this large black cloud flying over.

We were home on school holidays (May I think it was) and we knew about the testing going on at Onslow. Everyone knew about all the tests but we were not expecting to see the black cloud. It was supposed to blow away out to sea. Mother made us go inside from my memory. Father thought it would just fly away so stop worrying but mother was very worried because the big water tanks had open tops so that meant the water in the tanks could be affected if there was anything falling from the clouds. We girls were just leaving to fly down to Perth.

Mother was worried about our house water tanks and would the water become undrinkable. I remember mother writing to us at school later because she thought that the big black cloud would bring down the plane when we flew through it. I do not have that letter and Maureen cannot remember. The Big Cloud was suffocating. It felt like I was not able to breath. I remember Father was going out for the day and mother asked him not to leave the house but Father said "It will blow away soon." I still get nerves just thinking about how it turned the day into night. Not for long. Maybe about half an hour; maybe a bit more. I had breathing problems anyway and from memory it made my problems worse, each time. This happened twice while we were home on holidays. Everything went dark just like night. Even the birds were silent and the dogs went crazy each time. Everything went very still. I saw the black cloud first and called Mother to come and see. The big black cloud had turned day into night and the shape was like a mushroom sort of shape. In fact it happened twice, with the big black cloud flying over the homestead; not sure if it was 1954 or 1956.

One person who did find out what was happening was Stuart Stubbs. At the time of the atomic testing off Onslow, Stuart was the owner of the *Comet Gold Mine* just out of the town of Marble Bar. Stuart, his family and many of his workers lived at the Comet Gold Mine.

Below is an account by Stuart Stubbs taken from an interview that he gave to the "Western Australia Newspaper" many years later. I am not sure when this interview took place, possibly during the 1980s.

> "Stuart Stubbs told of how when he ran his Geiger counter over sheds, houses and machinery on the Comet Gold Mine after one of the tests, it went right off the scale; so he had his workmen hose everything down with water, then they retested the areas again and where radioactive material was still found, they again went over everything with the water hoses and continued to do this till Mr Stubbs felt that they had cleaned it all up or as much as they could. Mr Stubbs said when he reported his findings to the Government Authorities and told them of his concerns; he was advised to say nothing about what he had found." [57]

Stuart Stubbs and our father would often get together and have a talk when Father came in to Marble Bar from the station, so I assume they would have talked about what was happening in the area and on *Hillside Station*. I would guess that what Stuart Stubbs had seen and tried to clean up he would have talked it over with our father, but Stuart Stubbs would have also asked our father to keep what he had been told to himself. I believe our father would have spoken to Stuart about what he was seeing on *Hillside Station* and he would have asked Stuart for advice. I am unable to verify what took place between Stuart Stubbs and our father as it took place a long time ago and all the parties have since passed away. To my knowledge there has

---

[57] *The West Australian* Newspaper, 1980s

not been a medical study of the people who live in the North West during the atomic testing at Monte Bello.

I do remember Father telling Mother what Stuart Stubbs had said about the tests after we had gone into Marble Bar for a visit and also that Stuart asked Father if he had hosed the buildings down with water. Father thought it was a strange thing to ask.

My mother was quite worried after hearing what Stuart had said and did ask my father to wash the buildings down but Father was not as worried as Mother was, or even Stuart was, for that matter.

Thinking back I have often wonder why Stuart was so concerned to ask father about hosing the house. There must have been something to alert Stuart for him to hose down the building at Comet Mines

Figure 34: Comet Gold Mine, Marble Bar

In the mid to late 1950s father and other station owners began to notice disturbing changes that could not be explained in their stock – both sheep and cattle – that had not been there before. Father became aware that the cattle and sheep (at the time he still had sheep) were losing their condition even though there was ample feed around for them to graze on. They were not thriving as well as father thought they should and he found this quite strange and out of character at the time.

What also alarmed our father was the fact that he was losing both his pregnant cows and pregnant ewes. He started to notice that they were also aborting their foetuses late in the pregnancy; even more disturbing was the number of deformed offspring that the cows and ewes were dropping when they did go full term.

After watching this happen to his sheep and cattle, father started taking notes of everything that he observed happening on the station. This started out being only about the sheep and cattle, but after some time father widened his search to include information on the native animals on *Hillside Station* and found over time the same thing was happening with them.

Our father was shocked when he found that the native animals were showing the same problems as his sheep and cattle were exhibiting. The native animals were losing their condition and not breeding as they would have normally. Kangaroos were having deformed offspring. All species of birds as well as goannas, snakes and lizards were being affected as well.

Father, after having long talks with the other station owners and finding that they had the same problems as he had, came to the conclusion that the native animals as well as their livestock were suffering from a health issue unknown to them and the other station owners who lived in the area. The station owners, after finding out that they had all had the same problems on their properties, started to ask questions about the effects of the atomic testing being carried out at the Monte Bello Islands from 1952 to 1956 and wanting to know if what they were finding with their stock could be related to the atomic testing that had taken place.

At first our father and the other station owners put this change down to the result of the drought that the area was just starting to recover from, but then they noticed that the trees and general vegetation on their properties was also showing signs of stress and not growing as well or producing enough nutrients to feed the native animals or the station stock.

This state of affairs went on for some time, till finally the Australian Government was forced to listen to what the station owners were telling their government representative for that area.

I can remember my mother and father talking about all this when I was growing up and even after I had left the station; my mother thought the Government of the day had treated the station owners shabbily.

The Western Australian Agriculture Department, after viewing the information that the station owners had given them, were directed by the State Government to look into the situation. A scientist was sent to examine the affected areas. His instructions were to carry out studies on the ground water and rivers as well as the earth and native vegetation. He was also instructed to do studies on the native animals and stock including horses, sheep and cattle that were on the stations at that time.

The scientist's name was Dr Snook; our little sister Elizabeth had just given her puppy the name of Snooky so we were all very amused with the fact that Doctor Snook's name was similar to her puppy. I am not sure that Dr Snook shared our amusement at all and Elizabeth was not too impressed with the way the good doctor treated her and her little dog.

Dr Snook set up his operations on a station nearby and he was there for some time – possibly months – from what I can remember. He visited *Hillside Station* a number of times during this period but seemed reluctant to give our parents any information or helpful advice. I believe this also happened to other station owners as well.

As far as I know, no one ever heard what Doctor Snook's findings turned up, and he never offered any solutions to the problems that our parents faced at that time. From what I saw when he did come to *Hillside Station*, Doctor Snook had very little to say to our parents that was at all helpful to them in looking after their stock and running the station.

Father and mother waited and hoped that given time he would find an answer to their problems, but they waited in vain. I left *Hillside Station* in 1959 and – from my memory – the stock as well as the local flora and fauna never seemed to improve during our parents' time on *Hillside Station*.

Father was worried and so was Mother. She kept on saying that she felt that there was something wrong on the station. Our father agreed with her but they did not know what it could be. As I said, our father had talked to other station owners and found out that they were experiencing the same problems as he was on *Hillside Station* and they – like Father and Mother – had no idea what was going on.

What I have written about is mostly from what I can remember of that time and not from any scientific knowledge. The family did not keep any documents from *Hillside Station* that I could refer to after our parents died. I found nothing that could be used to verify or discount the above. Our parents never mentioned receiving correspondence from the Government of the day in regard to our father's concerns. The above is mostly hearsay.

It would appear now that the whole area may have been contaminated by the 1950s atomic testing that was carried out at the Monte Bello Islands.

It is interesting to note the fallout from the atomic testing in June of 1956 at the Monte Bello Islands (Trimouille Island G2; the yield was reported as 60kt, but was in fact 98kt). The fallout from this bomb contaminated an area from Onslow in Western

Australia, to Queensland which including Mount Isa, Julia Creek, Longreach and Rockhampton.

Effluent in the Shaw River

Father and Mother were also facing some other serious problems as their friend and neighbour Johnno Johnson had set up his mine operation too close to our main water supply – the Shaw River – which was a matter of yards from the homestead.

John Johnson had without thinking (at least I hope that was the case) set up a system so that when he was preparing the metal (tin) for sale they had to wash it a number of times before he felt it was clean enough to take to the Mines Department. He used the river for his water, and the problem was that he allowed the water that he used to clean the tin to run back down into the river. This waste water was nothing but contaminated sludge, and as he moved into lower grade ore the amount of sludge kept increasing.

Figure 35: Shaw River at Hillside Station (Photo Bree Krieger)

This ended up causing a lot of damage to the river and made the water unusable for domestic purpose as well as for the stock.

The result was that Father ended up taking Johnno Johnson to court, which was very unpleasant for both our father and no doubt for Johnno. Father won the case and the courts awarded him compensation for the damage done to the river, but it was many years before the river itself recovered from the damage that Johnno had done.

I was living in Perth during this time, so was not fully aware of what was taking place between Johnno Johnson and Father, but I understand Maureen may have more information as she was married and living in Marble Bar during this time if I remember correctly.

Stress and Strain

Father was in no state to take on more stress as he and our mother were still not coping at all well with loss of their son.

Being around our parents was not the most pleasant experience during this time in our lives, as both Mother and Father spent a great deal of their time yelling and screaming at each other and neither seemed too worried that there were children around who could see and hear what was happening.

Elizabeth was quite young and still at home during this period as she had not yet been sent away to Loreto. I can only image what it must have been like for her. All this yelling and screaming seemed to give our parents some relief from their own worries, at least for a while.

Our father turned to Maureen and Judy for his emotional comfort. This was to the detriment of Elizabeth. Mother also reached out to Judy and a lesser degree to Elizabeth for her emotional needs.

Problems soon started to emerge between Maureen and our mother. It must have been only a matter of months after our mother had returned to *Hillside Station* after little John Patrick died.

Mother was very unsettled and desperately unhappy over the loss of her child. She did not find the return to the station at all easy as the memories were just too raw for her to cope with. Father did not appear to offer Mother any support or comfort during these difficult times.

Maureen and Mother were both very strong willed people and Maureen was a feisty girl; a real survivor in our world. This resulted in Mother and Maureen clashing at every turn. Maureen questioned everything that Mother asked her to do. This of course drove our mother to distraction – which I rather think Maureen quite enjoyed at times – but things changed between them after our brother's death. Their conflicts appeared different from earlier times; it was as though Maureen and our mother were unable to reach out to each other after John Patrick's death.

Santa Gertrudis cattle

Father was interested in a breed of cattle being bred by the *King Ranch* in Texas which became known as Santa Gertrudis cattle. Our father had been keeping in touch with his friends in America who were in the cattle business, and I assume that is how he heard about the breeding program that Captain Richard King was conducting at his *King Ranch* in Texas, America. They were running a trial that involved the cross breeding of shorthorn cattle with their Brahman stock; they had started this cross breeding program in 1910, I understand.

Our father lived in America for quite a few years before returning to the family's *Austin Downs Station* at Cue in Western Australia, so it is also quite possible that he was familiar with or may have even worked on the *King Ranch* during his time in America.

Father had been following the progress of the Santa Gertrudis breed of cattle both in America and then here in Australia. The *King Ranch America* introduced the Santa Gertrudis breed of cattle into Australia in 1952. They established their head King

Ranch Australian quarters at Risdon, Warwick Queensland, and offered 12 bulls at a public auction on 14th November 1952.[58]

The King Ranch made a further importation in 1954. Sometime after this a total embargo was placed on importation of cattle by the Australian Government to prevent the possible introduction of blue-tongue disease. Importation of cattle stopped till 1981 when the Australian Government established a Quarantine Station at Cocos Island. [59]

Our Father was keeping an eye on what was happening on the *King Ranch* at Warwick in Queensland and was pleased with what he was hearing about these Santa Gertrudis cattle. He felt that they would be the breed of cattle most suited to the conditions on *Hillside Station,* if and when he made the move from sheep to cattle.

When the Government put an embargo on importing cattle from overseas, I believe it also included moving cattle within Australia. This could not have come at a worse time for our father because with the embargo that was then in place, it meant the end of Father's dream to bring Santa Gertrudis into Western Australia.

Our parents decided to still go with Father's plan, and changed from sheep to cattle which he bought from within Western Australia. This of course meant that Father would be gone for weeks at a time. If I remember correctly, he bought shorthorn cattle to be the main blood line for his breeding stock.

---

[58] When I went looking for Information on Santa Gertrudis I came across the Santa Gertrudis Breeder's Australia Association site – Santa Gertrudis Cattle Breed History. This is where I found information on the *King Ranch* in Texas America and also information on their Australian breeding program.

[59] The above information came from the Santa Gertrudis Breeders' Australia Association. Santa Gertrudis Cattle Breed History, page 1-2

Uncle Jack Meehan – our father's brother who ran the *Austin Downs* station out of Cue – came up to *Hillside Station* about this time. I can remember them having long talks about the benefits of introducing the Santa Gertrudis breed into Father's stock.

I believe that Father thought the best way to introduce the Santa Gertrudis blood line into Western Australia was by importing sperm from Queensland or better still from the *King Ranch* in Texas. That way they should not have a problem with the Blue-Tongue disease that had caused the Government embargo. This was of course a very expensive exercise and Father was not in a financial position at that time to carry out such an operation.

But after much talking between the brothers, Jack was also becoming interested in breeding the Santa Gertrudis for *Austin Downs*. Jack thought the idea of using sperm was worthwhile, so he and Father decided that Jack would begin investigating the possibility of using imported sperm from either Queensland or from America, which would of course have been dependent on the Australian Government of the day agreeing to his plan.

Jack offered to give Father a choice of bulls to take back to *Hillside Station* if Jack's plan to use sperm from Queensland or America bore fruit. This way both brothers would benefit from the semen that Jack was trying to bring into Western Australia. Father would then have been able to breed the Santa Gertrudis bulls with his shorthorn cows. This would be the most successful way – father thought – for his breeding program on *Hillside Station* to work.

Jack also offered to take over the debt that Father and Mother still had with Elder Smith (a Pastoral Company that station owners up north did business with). This loan would then be repaid once Father had the breeding program up and running. Jack told our parents that he would write this agreement into his Last Will and Testament. That way Jack could ensure that father did not lose the station if anything happened to his brother. Unfortunately Jack apparently never carried out that promise.

Jack as it turned out was unable (for a variety of reasons of which at that time Father and Mother were completely unaware) to carry out any of the promises that he had made to his brother Arthur when he visited him on *Hillside Station* in the 1950s.

It was only after Jack Meehan's death in 1962 that they became aware that *Austin Downs* station and Jack were facing some very serious financial problems.

Father then also found out that his brother Jack had never added the promised clause to his Last Will and Testament which would have protected our parents from losing *Hillside Station*.

Of course this all came as a dreadful shock to our parents, because if they had known what was happening on *Austin Downs* they would have been able to take some sort of action to save themselves as well as *Hillside Station*. They may even have sold *Hillside Station* and bought a farm down south or even retired.

Leaving Hillside Station

This was the last straw for our parents. In fact it could be said that it was the straw that broke the camel's back as in 1962 or 1963 they were forced to walk away from *Hillside Station* with just their personal items as the bank foreclosed.

They never got to fulfil their dream of having a Santa Gertrudis Stud which they would have been so proud of. Instead, Mother moved back down to Perth for the rest of her life.

Father took his riding horse and his pack horse and headed up to Wyndham where he found work in the cattle yards at the

Wyndham Meat Works. Father developed Parkinson' disease but continued to work there as long as he could.

He died ten years later, in 1973. Father never fully recovered from this disappointment and what he saw as a failure on his part. I believe he turned his back on all that was familiar to him, including his family.

Judy Returns to Hillside

After my sister Judy died in 2009, the family took Judy's ashes to *Hillside Station* in 2010 and placed some ashes on our brother's grave, and then they spread some ashes on the ranges that go past the homestead in memory of my family.

Santa Gertrudis Postscript

The interesting footnote to this story is that the people that bought *Hillside Station* from our father (the Stoneys – I think that was their name) bred Santa Gertrudis while they had the station. I think they sold *Hillside Station* in either 2014 or 2015.

Father would have been over the moon to know that the new owners – the Stoneys – like our father believed in the Santa Gertrudis cattle. Father had a dream and the Stoney family made it happen I understand. The Santa Gertrudis Stud would have made our father's dream come true. Father would have been so happy.

.oOo.

CHAPTER 19
Maureen's Story on Leaving *Hillside Station* in 1958

Below are two stories; they are the same story yet so different. They are about my older sister Maureen and how she was forced to leave *Hillside Station* in 1958 after she had left school and how in 1959 I was forced to leave *Hillside Station* in much the same way. The first story is Maureen's story.

Figure 36: Maureen Meehan

This is Maureen's story; and this is her story from what I have been told over the years. So bear with me. I have spoken to both Maureen and Elizabeth to have their permission to tell their story. This happened while I was away at school.

Once or twice a week our father would go over to the Shaw River Mine site to collect the mail bag that the plane would have left for station people. But getting the mail did not necessarily put our father in a good mood. When he collected the mail bag, he could be happy with the mail or he could be in a mad mood because of what was in the mail bag. It would be like walking on egg shells to keep both Father and Mother calm as the mail was not very often a happy event.

This always put the family on edge on mail day. Mother was on edge and so was Maureen. The family was ready for a fight and Maureen and Mother had been fighting since the day before anyway. Mother knew how to push Maureen's buttons and Maureen also knew how to push Mother's buttons; I do not even know if Mother and Maureen even could remember what they were fighting about.

When Father arrived home and found that his wife and his daughter were still fighting, he lost his thinking and was ready to join in. Now, when father thinks it is a good idea to join in with the two ladies who are fighting, then he needs to rethink his plan. Which he did not do. He joined in, taking Maureen's side against Mother. Mother was beyond mad. She was boiling. How dare her husband take Maureen's side. Maureen was stoked to have her father on her side. It did not matter what the fight was about by now; Maureen saw having Father on her side as winning.

Now Father and Maureen then thought that going to bed might cool every one down for the next day, but no one thought about the state of Mother and what was going through her head. Father did not think for one minute of the effect on Mother of him having taken Maureen's side and left a boiling implosion ready to go off. Maureen was still thinking that she had won and Father just thought that peace had settled, so I guess he would have gone to bed.

Maureen would have been chuffed still at having Father on her side and most likely would have gone to bed thinking that all was over and there was still tomorrow to sort everything out, and thank goodness that the yelling had finely stopped. Maureen would have gone to bed and not thought about Mother or what sort of state that Mother was in and assumed that she would be alright. After all, Father had gone to bed so from Maureen's thinking this must mean that Mother was alright. It was the next morning that the full results of the day before was known. Below is Maureen's story.

In 1958 Maureen found herself being removed from the station.

Mother and Maureen had one of their clashes which was quite a common event, but mother was still in a raging temper when she awoke the next morning. Mother was still in a yelling mood and still mad as hell by all accounts; mother was in a raging mad mood. My guess is that Maureen would have been done well to have gone out for the day with Father and avoided yelling for the rest of the day.

So when Mother decided to wake Maureen up she was still in a temper and not even willing to sit and try and talk things out with Maureen or Father. Mother was not in a good mood and raging at Maureen about all and sundries that she was boiled up about. When she arrived next to Maureen's bed the next morning, Mother was in a fighting state. Mother was getting ready to have a big fight; she was not in a friendly mood. Maureen went into shock when Mother awoke her with the demand that she had to leave *Hillside Station* that day by mid-day; Mother was not interesting in talking to anyone. No matter what anyone said, it did not matter. She wanted Maureen gone by mid-day, so the story goes.

Maureen at the time was living on *Hillside Station* after leaving Loreto at the end of 1957 and was not at the time sure what she would be doing. So, like me later, she was helping Father on the station and helping Mother by cooking and cleaning.

Our father did what he could to change Mother's mind but with no luck, so Maureen started to pack her belongings together. Maureen was still in shock and was starting to wonder what to do as the day went on. What to do, as she had no money and no job and nowhere to stay.

Father stepped in but could not change Mother's mind so he went into Marble Bar. Ian Blair was the first place that Father went. He told Ian what had been happening at *Hillside Station* and Ian spoke to Kath Pozzi. Then Ian went and spoke to the Mines Department to see if they could find work for Maureen. Kath Pozzi also spoke to them and told everyone how good Maureen was and how much she needed a job because her mother was not well and had not been very well for some time.

Kath offered to have Maureen board with her and the Mines Department offered to give Maureen a job. This sorted, Father returned to the station and told Mother what he had been able to find in Marble Bar for Maureen, and that now Maureen had a place to stay and as well that Maureen now had a job as well at the Mines Department.

I am not sure just who was the most surprised; Mother or Maureen. I think Mother was very surprised as she most likely did not think Maureen would leave.

Mother and Father were great at yelling at all that suited them. Maureen yelled at Mother, Mother yelled at Maureen and whoever else were around. This was always the way, no matter where we lived, be it *Hillside Station* or even in Perth. They all seemed to be yelling at all and whoever.

I tried yelling, but it was not my choice to be yelling all the time.

Maureen, I think, was glad to be leaving *Hillside Station*. She packed a few things and the following week father took her into Marble Bar, where she stayed for about a year.

The following year I left school too to stay at home too and I found having Maureen living in Marble Bar was great. I got to see Maureen much more often and because Father wanted to keep a watch on Maureen, Father also went into Marble Bar much more. I do not remember Mother ever going into Marble Bar with Father to see her.

Also, Father and Mother allowed me to go to the Marble Bar Races which for a teenager was very cool because Father let me buy a dress (much to Mother's anger).

Maureen was very stubborn; she had learnt how to survive. Maureen worked in Marble Bar for the Mines Department for about 12 months I think. Then she went away to Perth to try her hand at nursing. However, she got engaged to Don Stubbs, and had a wedding to plan, and Mother had moved to Perth and taken me with her.

Maureen came to some sort of arrangement with our father that he would allow Maureen to spend six months – I think – living on *Hillside Station* before she got married. Father very much approved of Don Stubbs, as did mother approve of Maureen and Don Stubbs getting married.

I am not sure what came next but Maureen returned to the station for about six months or a year till she left to marry Don Stubbs (Stuart and Alice's younger son); in 1961 I think it was.

So all were happy and it seemed to have worked for Maureen. Father taught Maureen all he knew and with that knowledge Maureen has been living in Bindoon on her farm.

This is my version of what happened to Maureen in 1958. My mother was a very tough lady. You did what Mother told you or you were in trouble. This is the story that I have been told over the years. I would like to think that one day Maureen will write her own version. I think she has a great story and I for one would love to hear her story. Mother cannot hurt any of us anymore.

.oOo.

CHAPTER 20
Virginia's Story on Leaving *Hillside Station* in 1959

This is my story of leaving *Hillside Station.*

I had finished school at the end of 1958 and was living with my parents on *Hillside Station,* helping our father on the property and Mother around the house, cooking and cleaning. Maureen was working in Marble Bar.

Then about the middle of 1959, father and mother told me that they had decided to go into Port Hedland for a few days as they wanted to catch up with some of their friends. I was told by my mother to pack a small bag as we were only going to be gone for a few days.

Figure 37: Virginia Meehan

A day or two later we left *Hillside Station.* I had packed a couple of changes of clothes and toiletries to take with me; everything else that was mine was left at the station. I had a half-finished book left by my bed, little personal treasures were still in drawers and my clothes were still in my wardrobe as I was expecting to return to the station in a matter of days.

It took us about two and a half days to reach Port Hedland as Father and Mother stopped along the way to visit with friends. On arrival in Port Hedland they booked me into the C.W.A. hostel (Country Women's Association) run by Vera and Norm

Burridge (I think that is how their name is spelled) for what was to be a couple of days, then they left to visit more friends. Luckily for me, Vera and Norm were very easy going, because my father left me with no money to pay for my accommodation and no money even to buy a meal so they ended up feeding me as well while my parents were out of town. I was very lucky in other ways as well. A lovely man took a liking to me and we went out on dates that included meals. He worked for Dalgetys from memory.

The days came and went, but Mother and Father did not return for over a week. Meanwhile I stayed on at the C.W.A. hostel in Port Hedland not knowing what was going on as there was no communication with my parents till my mother returned to Port Hedland and joined me at the C.W.A. hostel. From memory we must have stayed a further week while mother decided what she was going to do.

My questions about where was my father and why had we not gone back to *Hillside Station* with him were ignored by my mother. I did not learn that my father had returned to *Hillside Station* at the same time as my mother had arrived in Port Hedland till about a week after my mother had joined me. Mother then informed me that she was moving to Perth and that I was expected to go with her. She also told me that I would not be returning to the station again; not even to collect my things.

My father did not come to see me before he returned to *Hillside Station*. There was no letter or any other form of communication from him to explain what was going on. Meanwhile I had been left stranded in Port Hedland waiting for the two of them to come back. No goodbyes. Nothing.

I thought for years that I must have in some way been responsible for what took place between Mother and Father. He just removed himself from my life; it was to be some time before

I saw him again and, when we did see each other, he was cool and distant towards me.

Mother went on to tell me that I would be the main income provider and I would also be expected to take care of my mother and my sisters.

When we got to Perth I found that Mother had made plans to live in Cottesloe. She had also been to see Mr Ahern at the Aherns Store in Perth to see if he would give me some work – which he did. He was a lovely man.

After some time in Perth and while I worked at Ahern's Department Store, mother booked me into a business school, (Burroughs from memory). After some training there, I went on to find work in an office in West Perth (Western Australian Foods, I think it was called).

On pay day I was made to give my mother my wages for the week. The packet was not allowed to be opened by me. This was never spoken about and that went on till I left to get married in Victoria in 1968. My mother also vetted any men who were interested in me and opened any letters from any man who might write to me. My mother would open any mail that I received from men that liked me, and after going out with a man I was told to come to Mother and tell her all that had happened. To my knowledge I was the only member of the family that Mother treated like that.

When I told Mother that I was leaving Perth and moving to Melbourne the only comment that Mother said was that I could go if I wanted but I must keep on sending money home to her while I was away.

My mother and I did not fight; that was not my way. Mother was a difficult lady to deal with. You were not allowed to have an opinion or do what you wanted and she was quick to make threats which included stopping having contact with my sisters or other family members. This also included Mother trying to stop any contact with men that I liked or who liked me. Mother just told them I was not available.

Mother did approve of some men that I dated, but if Mother did not approve of a man she made her feelings known. Mother approved Tom Herzfeld and hoped that it would work out so she was nice to him. But the man from Scotland she was not happy about so did not support our friendship. Mother was pleased about Frank; he managed to charm my mother when he was in Perth. In 1973 I think it was.

Mother continued to find me work even if I was happy where I was working. Mother would arrive at where I was working and demand that she be allowed to go to lunch with me. That caused many upset bosses, needless to say.

I have written my version of leaving *Hillside Station* in 1959. That is all that I will be talking about what happened. It is over.

.oOo.

CHAPTER 21
Mother on the Move

Mother did not really cope very well living on *Mt Vernon*. The isolating life style did not sit well for my mother at all well. Mother could not make friends because other stations were some distance way from *Mt Vernon* so she had little social interaction. This meant my mother had moments when she became emotionally too fragile. Our father was quite different and could not understand why my mother was not settling down. He thought Mother should try harder but our mother found the alone was very hard. Then add to that the isolation which Father loved. The isolation and the alone, and our father found my mother's not coping hard to understand.

From the start of their married life, every so often Mother would pack up and go down to Perth for a few weeks. My father found it very hard to understand his wife's behaviour. All the other women had the same troubles and the other women seemed to be able to cope with the alone and the isolation.

Mother found it very hard to settle down in Perth when she left Father and Hillside Station in 1959 when I was 17. I think she had been in the bush too long and found city life rather suffocating. Anyway, Mother developed this habit of moving house. We would go to work or school in the morning, but that night on our return we would be told to pack our things because the next morning we would be moving house and suburb. Friendships suffered because of the constant changes of address. Boyfriends also fell by the wayside because of these moves. We seemed to be forever moving.

From the time that Mother arrived in Perth, she was on the move. A lot of the stress and problems that we faced while living with our mother in Perth came about because of her love of moving house. She would forget very quickly about the upheaval that all her moves caused the rest of the family.

Mother continued her nomadic life for many years. These are some of the suburbs that mother moved to. We had 5 different addresses in Cottesloe, 2 houses in Nedlands, 2 houses in Mt Lawley, 1 house in Como and 3 houses in Mosman Park from my memory. These were just a few of the moves that Mother made.

I think we must have lived in just about every suburb in Perth. Mt Lawley to Cottesloe, then to Como. Also Nedlands and of course Mosman Park. We went everywhere, which was very unsettling for Judy and Elizabeth. I just shrugged my shoulders and started packing. I think we became very well known to the removalists of Perth during this time of upheaval in our lives.

Mother's constant moves led to a very transient way of life for her and us girls. Unfortunately, this was to become our way of life for a number of years while my two younger sisters and I supported our mother and tried to take care of her. Finally the three of us left home.

Mother Returns to *Hillside Station*

Mother stayed on in Perth until Maureen married Don in 1961. Sometime after that she returned to *Hillside Station*. Apparently she and father had decided to try again to make their relationship work. Judy and Elizabeth were left in Perth as boarders at Loreto.

In 1962 Mother became ill on the station and was taken to the Port Hedland hospital for treatment. Father sent a telegram to tell me I had to return to *Hillside Station* as mother was in hospital and he needed my help. So I caught a plane back to the station and spent two days with father on *Hillside Station* before Don Stubbs came and took me into Marble Bar and I stayed with Maureen for a couple of days. Don then took me into Port Hedland to see mother. I was in Port Hedland for a day before I returned to Perth; it turned out that neither Father nor Mother had wanted to see me.

I boarded with a lovely lady in Cottesloe while Mother was back at *Hillside Station*. I was with her for a number of months till my Aunty Nell Partlon became most insistent that I move in with her and my cousins Geraldine and Mary. She did not like the thought that a family member was boarding with a stranger. I stayed with Aunty Nell till Mother returned to Perth after my parents' efforts to patch up their marriage failed when they lost *Hillside Station*.

Mother leaves *Hillside Station* forever

Figure 38: Maureen and Mother

It was late 1962 or early 1963 that Mother again moved back down to Perth. This time she declared it would be permanent, as they had lost *Hillside Station* and Father had gone away.

Life never seemed to run very smoothly after Mother came down from the station to live in Perth. My two younger sisters were still going to school at Loreto but as day students again. The girls were also growing into teenagers and wanted more freedom, which our mother had trouble coming to understand. So, of course, we had the odd blow up in the family.

When Mother finally returned to Perth, the problems that we faced before started again. So one weekend – in 1965 I think it was – I contacted our cousin Olivia who lived in Melbourne. I told her what I was planning, and with her support I booked my train ticket to Melbourne. With my fingers crossed, I told my mother I would be leaving in a week. Mother was surprised but she knew I had made up my mind so, apart from telling me that I would have to send money home, no more was said. Judy and Elizabeth came to the train station but not Mother. She did not even bother to say good bye and good luck.

Virginia goes to Victoria

Olivia Falconer was our father's cousin. Her father was J.P.'s brother, Will Meehan[60]. Olivia and her husband, Les, had been living in Perth, which is where I had come to know them. After Les died, Olivia had returned to Victoria to be closer to her sisters and her married daughters. I lived with Olivia for about two years in the 1960s in Melbourne.

Olivia had two sisters, Mary Meehan and Ellen Clemann, who were also very kind to me. Mary took me to meet their O'Rourke relatives[61] They were getting on in age but they were lovely and the farm was a dream.

Olivia, Mary and Ellen were so special ladies and I owe them so much for loving me and teaching and helping me. I knew Olivia longer since I already knew her from Perth before I moved to Melbourne where I lived with her for about two years. What a

---

[60] William Francis Meehan, who moved to WA with J.P. but returned to Cathkin, Victoria later in retirement. Will had a son Jack (John Patrick Meehan). This Jack's son was named William whose son became yet another John Patrick Meehan.

[61] Their father, Will Meehan, had married Helen O'Rourke of Eildon.

very special lady, and I have much to be thankful for that I met Olivia.

Mother and the Knife

As I understand from Judy and Elizabeth, things were pretty difficult at home, so Mother talked to Maureen and between them they asked Elizabeth if she would go up to Port Hedland and live with Maureen and Don for a while. Elizabeth had left Loreto by then and was working at an office in Perth.

I have one incident etched into my memory. It happened in 1966. I had been working and living in Melbourne Victoria since 1964 or 1965 and had returned to Perth for a couple of weeks to see my mother and sisters. I came back from Victoria for a week to see the family, and while I was there' Elizabeth and Mother had a big blow up. Mother had asked me to make Elizabeth some dresses (poor Liz), because she was moving back to Port Hedland

Elizabeth and Mother were having a humdinger of a fight and things were getting totally out of control.

Elizabeth would not give ground in their argument, and nor would Mother. Things went from bad to worse and finally escalated out of control. It was terrible to watch and Mother was becoming very physical towards Elizabeth and making threats about what she was going to do to her. At the same time Mother was shaking and manhandling Elizabeth.

I was unable to calm either of them down. Next thing I knew Mother had picked up a big kitchen knife and she was threatening to take Elizabeth outside and cut her throat.

In the end I managed to get between them both and talked Mother into giving me the knife. I managed to finally calm them both down. I believe to this day that had I not been there Mother would have carried out her threat to hurt Elizabeth. I think Elizabeth feels the same, as she has never forgotten that day. It had become a very volatile situation for all of us

Figure 39: Elizabeth Meehan and Brian Casey

How much this has hurt Elizabeth and affected her life is anyone's guess; but I can still see the fear in her eyes as she must have thought she was going to die within minutes if I could not calm Mother down.

Elizabeth took great delight in winding up our mother. Just as Mother would take great delight in winding up Maureen. But Elizabeth was a good daughter and became the main breadwinner for herself and Mother after I left Perth for good. I have often wondered if Mother and Father ever fully appreciated what Elizabeth did for them. Or how much Maureen and Judy did to help and support them in their later years.

Judy was the only daughter that mother seemed to listen to and they appeared to get on very well together while Elizabeth was away in Port Hedland.

I finally returned to Perth in late 1966. Mother was living in Cottesloe at that time and my return did not really work out. Mother was back moving house every full moon, and there were still the constant fights she had with Elizabeth, and with Maureen when she was down visiting from Port Hedland.

While living in Melbourne, I met Frank O'Shannassy. We were married on the 26th of October 1968 in Melbourne. It was a small wedding, but Olivia, Mary and Ellen were there to support me. Ellen's husband, Ossie Clemann, gave me away. He was a policeman and a lovely man. They lived at Lilydale and four adult children plus many grandchildren at that time.

Figure 40: Judy in Pilbara landscape

My trips to Perth became less and less over the years. I brought our daughter Marita over when she was about fifteen months old, then Frank and I came over again when Marita was about two years old. The three of us also came over for Elizabeth and Han's wedding in 1977.

Mother died in Perth in 1977, four years after Father's death. She was 67 years old at that time.

.oOo.

CHAPTER 22
Arthur and Kitty

When I started writing this story, I was filled with a sense of adventure and foreboding. Foreboding about what I might learn after all these years – both the good and the bad memories that would automatically surface – as I worked on this story.

Our father loved to tease our mother by telling her that he has seen a friend of hers or he would ask Mother 'had she heard from Doctor John'. Mother did not blush but over the years she told us that was how Mother and Father met. Mother was nursing at the Meekatharra hospital and John was a doctor there.

The story that Mother told me (and told my sisters as well, I would guess) was that Grandfather J.P., and Grandmother Minnie held a big party at *Austin Downs* to celebrate our father's return to Australia from America. Father had been living in America for about five years or more and Minnie and J.P. wanted to give a big welcome party weekend.

Apparently, it was more Grandmother than Grandfather behind the party, as J.P. had wanted a small party. Mother said that all the doctors and nurses were invited, as were all the young and pretty girls and handsome young men of the district and businesses. Mother said she was not too keen about going but Doctor John said it would be fun, so off they went.

It was getting late on Saturday when the door opened and this dashing, handsome man came in. My mother said it was like a bright light took over the room. She said that the young ladies were all over my father. The girls were hanging all over Arthur and he just stood there. Mind you, the men were also in awe of this man who had a strange accent. Some said it was like a little of this and a little of that. Australian, some American and a touch of an unknown accent (Arthur said it was Mexican). What an impression it made and what charm this man had. He was quite happy moving around the room talking to his mother and father and also to friends from when he lived there.

The media talk about charisma. Well, from what my mother said, Arthur had it in bucket-loads. We grew up seeing and hearing Father's stories and seeing how people loved listening to him talking about his time in America. He loved talking to people and people loved him. We saw how our father could keep people enthralled. He just seemed to keep people wanting to hear more stories. He had a way of telling stories and he loved talking about his time in America.

Mother said it was a sight to behold but she also said that he stood out from the other men at the party. J.P. got on well with Kitty and was trying to get them talking together. Arthur said he liked Kitty but he thought she was shy. Kitty said Arthur was very good looking but she liked Doctor John; he was a gentleman. It was some time after the party that Arthur and Kitty fell in love and were married.

Mother had never been to *Mt Vernon* and had just accepted what Father said about the homestead. It was a shock, seeing the homestead after the wedding. Coming from a close family, and with close nursing friends, Mother was not ready for Mt Vernon and what it would mean for her, and how much she would have to change (if she could even do that). Mother was out of her comfort zone and missed Doctor John very much.

Mother was frightened by everything and did not quite know what to do first. Mother was not happy only being able to do her shopping with a Pedal Radio. Poor Arthur. Poor Kitty. They were not suitable for these two lovers who fell in love.

Father was an action man doing cattle things and mustering and riding horses. Mother was into reading books and having morning tea with her nurses and friends; anything but trying to cook on a wood stove in the kitchen. Mother was lost and Father was not far behind. Father said that he just thought Mother would fit in and not be frightened or nervous.

Grandmother Murphy, from what our mother and her sisters told us, was a very sweet lady but she lived in fear that she or her children might upset her husband, William Patrick Murphy, who could then turn on her and the children as he had a violent temper, and was easily aroused. The children were not allowed to talk at the dinner table and if they did they were taken outside and punished. Aunty Meg was always in trouble for talking at the table, from what our mother told me.

Grandmother Murphy and the girls would find themselves the subject of his temper quite often, as he would then punish both his wife and the children for whatever perceived misdeed that he believed they had committed. Even Thomas (Tom), the youngest child and the only boy, was not safe from his father's rages.

As children we saw the results of Grandfather Murphy's violence surface over and over again in the way that our mother handled emotional and physical pain; she appeared to always internalise her feelings.

Father on the other hand and because of his upbringing seemed to be unable to show his love, especially when he was under pressure. I have no doubt that they loved each other very much, but our father did not seem to understand that love and support were an integral part of any relationship. He always seemed to find it difficult to give his love and support to Mother when she needed it the most.

The one solid thread I found that seemed to connect both our father and mother was their similar family dynamics. Both our parents had very tough upbringings; their fathers were hard men who believed in not sparing the rod when it came to disciplining their children. That was certainly the case with Grandfather Murphy and his children carried the scars from his draconian ways for the rest of their lives. Our mother certainly carried some horrific memories from her childhood.

The legacy that Father and Mother inherited from their parents proved to be very toxic. Even as children we were made very aware of this in the way that our mother struggled with the events in her life, and also in the way that she coped with the tragedies which she had to endure while living on *Mt Vernon* and *Hillside* stations.

Mother and Father tried very hard to make their marriage work. Father loved mother and Mother thought the world of Father but events were too strong for both and what went wrong with the death of their two babies was too much. Mother suffered greatly after her babies died and the second death of John Patrick was just too much for both of them. They could not pick up the broken pieces and move on, which was what my father was trying to do. It was the only way my father could handle everything. This meant that Father could not take care of us, his children.

Mother was no better, so she tried to send Maureen away. Mother then wanted to get rid of me and my father could not understand that he needed to save Maureen and me from Mother so off I went too. Father could not even try and find a home in Marble Bar for me.

It would be easy to just say that my Father did not care about me, but I think it was even worse. Father was out of petrol. He did not have any energy and no love to fight over. As mother had died inside of her heart so had my father. They both loved us all but they could not help any one of us anymore.

We love you, Father and Mother; you did your best.

.o0o.

Appendix 1 – McCarthy Family

J.P. Meehan's wife was Minnie (Ida May Blanche) McCarthy. Our grandmother was the daughter of Denis McCarthy and Clementina Langoulant.

Denis McCarthy (1828 – 1888)

Minnie's father, Denis Domenic McCarthy, was born about 1828 in Donoughmore, Cork Ireland[62]. On the 14th of February 1847 (when he was 19 years old) he married Julia Dun in Newmarket, Cork but he was convicted and transported to WA in 1850 aboard the *SS Phoebe Dunbar*. Dennis applied to have Julia join him, but she defaulted and stayed in Ireland. They never met again.

Denis was about 41 years old and working as a plasterer in Perth when he married Clementina Anna Augusta Langoulant. The family always said Denis was from an old and well established Northam family and that he was 21 at the time of his marriage.

I have been able to get copies of some of the family's death certificates as well as a Wedding Certificate for Clementina.

Clementina Langoulant (1851 – 1920)

Clementina Anna Augusta Langoulant was the 4th surviving daughter of Mary Ann King and Louis Langoulant. She was born probably in their hut in Wellington Street, in what is now the Perth city centre. Her father was French and well educated, and it is said that they all had to speak French at home.

She told people that she was 21 years old at the time that she got married, but on her death certificates it says that she was only sixteen. I guess that her father, Louis Langoulant, would

---

http://gw.geneanet.org/podonnell1?lang=en&pz=patricia+anne&nz=langoulant&ocz=0&p=dennis+domenic&n=mccarthy

have given any information regarding his daughter Clementina, as well as the wrong information of her age. She was working as a domestic when she married, but with lots of younger brothers and sisters she was no doubt used to that.

Clementina and Denis McCarthy married in Fremantle at the Congregational Chapel on the 16th of September 1869. (This was unusual as Clementina had been raised a Catholic, and later also raised her children as Catholics I tis probably that Denis was also a Catholic).

Denis and Clementina went to live in Northam, where they raised eight children and also lost one at birth. They were:

Ellen (Helen Victoria) McCarthy (1870 – 1952)

Ellen married Mr J.E. Hicks and died aged about 82 on the 29th of December 1952. They had three sons (Roy, Fred and Arthur Hicks) and a daughter, Florrie Hicks Attridge.

Hannah (Hannah Clementina) McCarthy (1873 –)

Hannah was provisionally appointed to teach at the Clackline Government School in 1896 when it first opened. She was then 23 years old. She was replaced at the school in May 1899. Hannah went on to marry Mr C. Hansen and lived at Southern Brook.

Arthur Dennis Joseph McCarthy (1876 – 1898)

Arthur McCarthy would have been 12 when his father died. He himself died in a railway accident at York on the 15th of June 1898. He 'sustained terrible injuries' including a fractured skull and broken spine and died that afternoon in hospital.[63] He was 22 years old and single.[64] This was the uncle for whom our father was named.

Minnie (Ida May Blanche) McCarthy (1879 – 1936)

Our grandmother (Minnie) who married J.P. Meehan.

---

[63] *Eastern District Chronicle (York WA)*, 25 June 1898, page 3: The Recent Green Hills Railway Accident

[64] *Eastern District Chronicle (York WA)*, 16 July 1898, page 3: Benefit Concert.

Gracie (Victoria Grace Marian) McCarthy (1881 –)

Gracie married William J. Morrison in 1912 when she was 31 years old. They lived at Northam.

Percy (Henry Louis Percy Langoulant) McCarthy

Born in 1883. Henry moved north to Meekatharra and spent most of his adult life there.

Josephina Clementina McCarthy (1886 –)

Josephine (Mrs Duffield) had a daughter, Pearl Duffield. Pearl was a bridesmaid for her Aunt Gracie in 1912 and attended St Joseph's School in Northam in 1917. They later lived in Brisbane.

Ivanhoe Ernest Lionel Vivian McCarthy (1888 – 1915)

Ivanhoe married Alice Lyons in 1914 before he enlisted with the 10th Light Horse on the 26th of October 1914, at Blackboy Hill in Western Australia. He was ill in hospital in Egypt in August of 1915[65], and was injured twice[66] before he died as part of the 32nd Battalion at Celtic Wood, Ypres, France on October 13, 1917. He was 27 years old and left no children.

Eveline McCarthy

Eveline died as a baby but I am not sure where she fitted in the family.

Denis worked as a bricklayer and plasterer but had frequent problems with alcohol and appears to have been a regular in the police courts. By 1884, things were not going well and Clementina charged Denis with assault, for which he got 1 month hard labour.[67]

Clementina supported the family as a seamstress and went to work at the Northam Hospital. Things went well for some years till Clementina met with a most unfortunate accident. Because

---

[65] *Western Mail*, 27 August 1915, page 1 (photo of Ivanhoe)

[66] *The Northam Advertiser*, 24 January 1917, page 2: Local and General News "Mrs W.J. Morison has received advice that her brother Ivanhoe McCarthy…has been wounded twice and ill on two occasions…"

[67] *Police Gazette Western Australia*, No. 35:1884, page 144

of the type of work that Clementina was doing, her eyes started to give her trouble so she went to the local chemist for some eye drops that she could use to ease the discomfort. Unfortunately for Clementina, the chemist gave her the wrong drops and they left her partially blind.

Denis died on the 16th of June 1888, aged 60 years, when he fell from the roof of the cottage he was building at Northam. A relief fund was set up to help Clementina at that time as "his widow is blind and has, we understand, no less than six children dependent upon her". £5 was raised.[68] At this stage Helen (the eldest) was 17 years old and Ivanhoe (the youngest) was a baby but Helen and 15 year old Hannah were considered old enough to be independent. Minnie was 9 years old when her father died.

I also understand that the hospital gave Clementina cleaning duties in place of the sewing that she had previously done but was now unable to do. In this way Clementina was able to continue supporting her family with the aid of the older girls.

In 1896, William Patrick Meehan (our great-grandfather) was dying in Northam and was apparently boarding with the McCarthy family. Possibly the hospital sent him there as a way to help Clementina with a little more income.

At this time, Hannah was teaching at the Clackline School and her pupils included her sister Josephine and brother Ivanhoe. Prizes were awarded to 21 students as reported in the newspaper report.

> "On Tuesday last the annual distribution of prizes to the pupils attending the Clackline Government School was made by Mr. J. T. Reilly. Prior to distributing the prizes, Mr.Reilly heard the children read and examined them in grammar, arithmetic and geography, and at the conclusion of the examination made a few remarks encouraging the children to pursue their studies, and was also complimentary to the

---

[68] *The Daily News (Perth WA)*, 18 June 1988, page 3: The McCarthy relief Fund.

Mistress, Miss H. C. J. McCarthy.... Prizes were awarded to the following children:-- Josephine McCarthy ... Ivanhoe McCarthy..."[69]

Then in 1898 disaster struck the family. The elder son (Arthur Denis Joseph McCarthy) was working as a guard on the ballast train while building the York to Greenhills railway and died when trucks were derailed. The train was backing up, and Arthur was sitting in the 'front' truck to guide the way. The truck apparently hit stones on the rails at a road crossing, rolled and crushed him[70]. Arthur was 19 years old and supporting his mother, so the people of the district ran a concert and dance for Clementina and the children. This time £10 was raised to help them.[71]

Sometime about this time, Minnie was learning the nursing trade, possibly in the Northam hospital. By 1902, four years later, Minnie had met and married J.P. and moved to the North West. She named her first son (our father) after her lost big brother Arthur.

Their brother Percy (Henry Louis Percy McCarthy) also went north and settled in the Meekatharra area. The remaining sisters also all married. They were Hannah (Mrs Hansen at Southern Brook), Gracie (Mrs Morrison at Northam), Ellen (Mrs Hicks in Perth) and Josephine (Mrs Duffield in Brisbane).

Their youngest brother, Sergeant Ivanhoe Ernest Lionel Vivian McCarthy, had died in France in 1917 during World War I.

Ivanhoe's two cousins (the sons of his Aunt Josephine McCarthy Duffield and Aunt Ellen McCarthy Hicks) also both died

---

[69] *The Northam Advertiser*, 19 December 1896, page 2: The Clackline School.

[70] *Eastern District Chronicle (York WA)*, 25 June 1898: The Recent Green Hills Railway Accident

[71] *Eastern District Chronicle (York WA)*, 16 July 1898, page 3: Benefit Concert.

overseas in the First World War[72]. I will keep trying to find out more information on Clementina's soldier nephews, but not knowing the names of our great uncles is proving to be a challenge.

Sometime after J.P. and Minnie were married in 1902 and Ivanhoe was out working, Clementina move north to live at *Austin Downs*.

Despite her lack of sight, Clementina still made the journey south to spend time with family and friends occasionally, and it was during a stay at Southern Brook with her daughter Hannah Hansen that she met her fatal accident.

When I was researching J.P. Meehan, I sent away for copies of marriage certificates and also copies of the birth and death and wedding certificates of Clementina and Denis McCarthy (J.P.'s parents-in-law). Below I have written the information that I have found from death and wedding certificates.

The death certificate of Clementina had quite a lot of information. On Clementina's Death Certificate it said that she had some sort of accident with either a fire or something hot (maybe a cooking accident) which meant that Clementina suffered from burns and died from shock after some hours, in the Northam Hospital. Before she died (11th March 1920, aged 69 years) Clementina had problems with her eyes so that may have been why she had the accident.[73]

These accounts from the local newspapers of the time give an account of what happened.

---

[72] The above information was gained from the Wikipedia the free encyclopaedia

[73] The above information came from the death certificates and wedding certificates that I was able to get copies from the Registry of Births, Deaths and Marriages' of Western Australia.

"Yesterday morning at the residence of Mr. C Hansen, Southern Brook, Mrs McCarthy, mother of Mrs Hansen, accidently got her clothes alight, and was severely burnt about the legs, and is suffering from shock. Dr Aberdeen was immediately summoned by Mr Hansen, and the doctor motored to Southern Brook, and attended to the sufferer. He ordered her removal to the Northam Public Hospital, where she was conveyed by the Ambulance, The old lady was 70 years of age and is reported to be in a low condition."[74]

Despite all the efforts, Clementina could not be saved

"The death occurred at Northam on last Thursday week of Mrs. Clementina McCarthy, relict of the late Denis McCarthy, one of the pioneer contractors of Northam, as the result of a burning accident. The deceased lady was on a visit to Northam from *Austin Downs* Station, Day Dawn, where her daughter, Mrs. J. P. Meehan resides. She was 70 years of age, but although her brief illness was necessarily a very painful, one, she was conscious till the end, having been fortified by all the rites of Holy Church. The remains rested in St. Joseph's Church, Northam, on Friday night. Requiem Mass was celebrated by Father O'Donnell on Saturday morning, and the funeral took place at 1 o'clock, which was very largely attended."[75]

My father Arthur (17) and his brother (our Uncle Jack, then 15) attended their grandmother's funeral along with J.P. and Minnie.

Clementina's death certificate was dated the 6th of July 1920 and gave a list of her children with Denis McCarthy who were still alive at that date. Below is the list of the children that were named there.

Helen Victoria 49 years – (Ellen Hicks)

Hannah Clementina 47 years – (Hanna Hansen)

Ida May Blanche 41 years – (Minnie Meehan)

---

[74] *Goomalling-Dowerin Mail*, 12 March 1920, page 4

[75] *W.A. Record (Perth WA)*, 20 March 1920, page 12

Victoria Grace Marian 39 years – (Gracie Morrison)

Henry Louis Percy 37 years – (Percy at Meekatharra)

Josephina Clementina 34 years – (Josephina Duffield)

The death certificate also shows that there were two boys deceased and a girl deceased. These would have been Arthur, Ivanhoe and baby Eveline. An Ellen is also mentioned but this was the eldest daughter, Helen, who was mostly known as Ellen. It is said that Clementina also had a daughter Pearlie but it may be that this is Pearl Duffield , whose mother was Josephine Duffield (nee McCarthy). Thus the Pearl mentioned was probably Clementina's niece and not another sister.

Clementina's Death Certificate had her father Jean Charles Langoulant (known as Louis) as a landowner who also worked as a labourer, and her mother Mary Ann King as home duties.

See Appendix 2 for the King Family Story and Appendix 3 for the Langoulant story.

The Black Sheep of the Family

To add some intrigue to the family story I have included some good old fashioned family gossip about one of our Great Aunts (one of Clementine's daughters). I have not learnt the name of this daughter so I will call her XXX. I have no way of checking this story for its authenticity so I would suggest that it be taken with a large grain of salt, but it is a good fun story for the family to enjoy.

Clementine's daughter, I have heard it said, got into trouble with the Law. From the story I was told, she became involved with a rather shady character when she was quite young and one thing lead to another and together they carried out a robbery on a bank or store in Kalgoorlie or York. Needless to say, they were caught, and it is said that XXX spent some time as her Majesty's unwilling guest.

Now being a rather strong willed girl and feeling no ill effects from her first little misadventure, she decided to head for the bright lights of Perth. Being some-what short on life skills and work experience, it took XXX some time to find work but she finally found a job and settled down, in Perth I believe. I am told that XXX finally married an Italian man and they settled in York.

As I said at the start this story is just gossip and should be taken with a pinch of salt. I believe my Great Aunt would find these stories quite amusing.

From memory, our father and mother always referred to XXX as the Black Sheep of the family – Father with a twinkle in his eye. Both Mother and Father's families had one or two black sheep of their own, especially the Irish connection on our mother's side.

.o0o.

Appendix 2 – King Family

Clementina Langoulant McCarthy's mother was Mary Ann Lucilla King (1825-1886)

Mary Ann was born in 1825 in Redlynch, Wiltshire, England. She had arrived on the barque *Calista* on the 5th of August 1829, aged 4 years, with her parents David King and Harriett Sophia Newman.

They landed at Fremantle and were rowed up the Swan River and landed under Mt Eliza (near to the old Brewery site), where Mary Ann was handed ashore to become, according to family lore, the second 'white woman' to step foot in the colony[76]. Her 8-month pregnant mother probably had a harder time getting ashore.

Mary Ann had a big brother (Henry King, aged 6) and their sister Sophia Harriett was born just a month after they arrived.[77]

Sadly, seven-year old Henry drowned in 1830. His father, David King, died on the 7th of February 1830, aged 34 years, just 6 months after their arrival. Maybe he died trying to save his son.

With two small children to care for, their mother re-married the following year to John Stanton[78] who was born in Ballina, Mayot, Ireland in 1794. He was then 36 and Harriett was 27 years old; they were to have just one child together (Emma Eliza, who married Hugh McVee and died aged 28 in 1865).[79]

---

[76] *The West Australian*, 19 December 1944: Looking Backward. An Eightieth Birthday

[77] Sophia is listed as the second white girl born in the new colony. In 1849 she married James McNamara.

[78] John Stanton had arrived on the HMS Sulphur as a private in the 63rd Regiment. He discharged and joined the Police Force.

[79] Brady family Tree https://www.bradyfamilytree.org

John Stanton apparently treated his step-daughters well, even after their mother died and he remarried Frances Savage. He gave Mary Ann a block of land on Wellington St in central Perth (next to what became the Globe Hotel) as a wedding present when she married Louis Langoulant. This gave her some security but also became an issue for their descendants many years later after Mary Ann's death in Perth on 31 July 1886 aged 60 years.

Mary Ann King Langoulant

Our great-great-grandmother, Mary Ann grew up in the earliest days of the settlement on the Swan River. She was married to Louis Langoulant on the 4th of January 1842 when she was 17 years old and he was 28. She had her first baby before the year was out, and went on to have at least 12 children. She also appears to have maybe had twins. Her last child was born when she was 44.

Mary Ann had been raised as a Protestant but joined the Catholic Church three years after their wedding. (Louis was still 'Charles'.)

> "Charles (Langoulant) plies the fisherman's trade. He employs two Frenchmen, one of them having ... been raised a Calvinist. The other... is a good Catholic. The wife whom Charles married three years ago was also a Protestant. Both these persons have had the happiness of coming home to the bosom of the Catholic Church. It was Dom Serra who baptised both of them yesterday, Monday. They are the two first fruits of the mission and so things were done with some solemnity ...The Sisters of Mercy, who did the instructing of Charles' wife, set great hopes on her fervour... Never could a man have met a better wife than the good Charles has done."[80]

In addition to her own children, she also took in the four children of her eldest daughter and raised them too after Ann died.

---

[80] The Correspondence of Leandre Fonteinne. Edited by Peter Gilet. P162

In 1881, when her youngest son was 12 years old, Mary Ann was involved in a serious accident which eventually ended her life at the age of 57.

> "An unfortunate accident occurred last Wednesday afternoon. An old woman named Langoulant, the wife of a lime burner, of that name, was crossing the railway line, with a cart, near the half-way house on the Fremantle road, just as a down train from Perth was approaching. She got through safely, but returned to shut the gates, and, being very deaf, did not hear the train, although the engine whistle was loudly blown. Every effort was made to stop, but the driver was unable to do so in time, and the poor old creature was knocked down. She was taken in the train to Fremantle, and conveyed to the hospital, where she was found to have sustained serious injuries about the head. That she was not killed on the spot seems almost miraculous."[81]

Mary Ann died some time later, in 1886, of the injuries received.

.oOo.

---

[81] *The West Australian*, 9 December 1881, page 2: Accident

Appendix 3 – Louis Langoulant

Our great-great-grandfather, Jean Charles Louis Langoulant,[82] was born in 1814 in France. He spent some of his childhood in Cherbourg, Manche, in Basse-Normandie and then ten years in Paris before he went to sea. As a sailor he then travelled:

> " ...all the coasts of Europe, of Greece and the Orient, America and the ports of Africa... the islands of Oceania and all around New Holland."[83]

Louis had a sister Jeanne Victoire Langoulant who was 5 years his senior and the 'superior at the civil hospice at Cherbourg'. Louis himself was apparently deeply religious and kept in touch with her when he could. He came to the French whaling station near Esperance in about 1838 on the *L'Harmonie*.

> "Whalers, mainly Basques, entered the Southern Seas in the 1830s and operated off New Zealand and Tasmania particularly but also along the south-western coast of Australia where there was a small French community of whalers near Esperance. Some deserted their ships and tried a new life on shore, such as Louis Langoulant whose descendants still live in Western Australia."[84]

Louis is said to have walked to Perth with his friend Charles Francois Tondut after they 'jumped ship' in Albany from the whaling ship *L'Harmonie*. When he first arrived in the Swan River colony, Louis described himself as an art/jeweller but he worked as a fisherman to keep himself. He was living on the banks of the Swan River in a hut at Freshwater Bay at the time

---

[82] He used each of his fore-names at various times in Australia, but later seems to have settled on 'Louis' so that is used in this book.

[83] The Correspondence of Leandre Fonteinne. Edited by Peter Gilet. P511

[84] The Australian People: An Encyclopaedia of the Nation, Its People and Their Origins. Edited by James Jupp. P356

and was claimed as a good friend by French Catholic missionary Leandre Fonteinne.[85]

Louis Langoulant and Mary Ann King were married on 4 January 1842 in an Anglican ceremony in Perth[86] when she was 17 and Louis was 28 years old.

Mary Ann's stepfather gave her Lot V19 on Wellington Street in Perth as a marriage settlement to be jointly owned during their lifetimes and then to pass equally to surviving children. In due course Louis built a dwelling there, though he also maintained the fishing hut.

In 1849 Louis became a British citizen and they had three daughters aged 7, 5 and 1 (Ann, Emma, and Florentine) and had lost Catherine as a baby. Louis tried to fit in time to educate all his children to the standards he required, including insisting that they all spoke French as well as English.

He earned his living in various small businesses, including running a 'passage boat' service between Perth and Fremantle.[87] This was a '4 oared row boat' and he was rowing it himself. He also tried to get a windmill operational and ran a bullock team at various times.

On the 7th of October 1852 Louis sailed for Adelaide on the barque *Merope* on his way to the Bendigo goldfields which had just been discovered. He left his wife with Ann (10), Emma (8), Florentine (4), Clementina (1) and she was also pregnant with Maude who was born the following year. By this time apparently they were living on the Wellington Street block and as a younger son Henry recalled:

> "...The Esplanade, Murray and Hay Streets had made roads. Our Wellington Street was just a track though the dirty

---

[85] The Correspondence of Leandre Fonteinne. Edited by Peter Gilet. P511

[86] http://gw.geneanet.org/podonnell1?n=langoulant&oc=&p=jean+charles+louis

[87] *Inquirer (Perth)*, 20 November 1850, page 4

black sand which fringed the swamp where he and his pals
caught gilgies and where the Central railway station now
stands."[88]

Louis was away for about six years, and Matilda was born in
1859 to mark his return. The following year baby Annie lived
only for a month. Just a year later, Mary Ann gave birth to
John.

By 1862 the family had 40 acres at Swanbourne (originally
called Butler's Swamp) and started building a farmhouse of
local limestone and she-oak timber called 'Pleasant Valley'
which had a ship porthole in a front wall[89]. It was in a sheltered
valley behind the sand-dunes along the beach with a productive
vineyard where Louis made his own wine. Their eldest daughter
Ann had married and left home but they had four daughters
(Emma (18), Clementina (11), Maude (9), and Matilda (3) and
their baby son, John. Florentine had married so there were five
at home in Wellington Street.

In 1866 Louis held the rights for getting wood and stone upon
the reserved lands of the Perth City Council[90] and operated a
lime-burning operation. Between 1852 and 1882 he employed 46
Ticket of Leave men on various occasions, mostly as lime-
burners, sawyers and gardeners.[91]

The new house was completed by 1867, but tragedy struck when
their eldest daughter Ann died at only 25 years, leaving four
small children under 6. Their father (Richard Keegan) couldn't
cope with them, so they were added to the Langoulant
household. By this time Louis and Mary Ann had added two

---

[88] *The West Australian*, 19 December 1944: Looking Backward. An Eightieth
Birthday

[89] The 'Pleasant Valley' house was demolished by council order in 1959.
(http://www.wslg.wa.gov.au/Clarelibweb)

[90] *The Inquirer and Commercial News*, 1 August 1866, page 3: Perth Police
Court.

[91] www.friendsofbattyelibrary.org.au/PDF/Dictionary%20of%20WA/L.pdf

more children of their own (Henry and Mary Ann II) but Emma had married leaving Clementina and Maude to help with the eight smaller children (including 2 toddlers and two babies).

Mary Ann had one last baby – a son called Joseph – in 1869 when she was 44 years old. This was the year when her daughter Clementine (our great grandmother) married Denis McCarthy. She left her mother with Maude (then 16) to help manage the five remaining children at home plus the four grandchildren. They were Matilda (10), John (8), Henry (5) Mary Ann II (2) and baby Joe plus Thomas Keegan (7), Mary Ann Keegan (6), Emily Keegan (4) and Ellen Keegan (2).

In 1880, the family lost 13 year old Mary Ann (II) when she died after falling out of a tree. Then in July of the following year, one of the boys had a nasty accident when his horse shied at a train.

> "A lad named Louis Langoulant, son of a lime-burner who lives near the Halfway House, met with a very severe accident last Sunday week. It appears that he was riding along the road, and when the train from Fremantle passed him he was thrown violently against the railway fence. How the accident exactly occurred we do not know. Some say that Langoulant was trying to race the train and that his horse unexpectedly fell with him, while others assert that the lad was not racing at all but was riding along at any ordinary canter and his horse suddenly shied violently at the passing train and unseated the rider. The lad was removed to the Perth Hospital where he lies in a critical state. It has since been reported that he had succumbed to his injuries, but the rumour lacks confirmation."[92]

John at this stage was 20 years old, Henry was 17 and Joe was 12 years old. I have not been able to work out which it was.

---

[92] *Herald (Fremantle)*, 16 July 1881, page 2

Then in early December 1881, only 5 months later, Mary Ann (their mother), was hit by a train when crossing the line near their home.[93]

At that stage the unmarried family were John (19), Henry (16), and Joseph (11) plus three of the Keegans: Thomas (18), Emily (15) and Ellen (14). Sadly, Mary Ann did not recover and died of her injuries in the home of her married daughter Matilda Whitfield in King Street in 1886.

In 1886 Louis Langoulant (then 72 years old) married Hannah (Annie) Rodgers who was 20 years old. Their first baby was born the following year, when Joe (the youngest of his first family) was about 17 years old.

Unfortunately, it was not a happy marriage. In 1889 Louis rented out and then sold the *"Pleasant Valley"* property[94] including 'vines in full bearing' and moved to the block in Wellington Street where they ran a boarding house but Annie was discontented. She apparently wanted the boarding house and land signed over to her, but had not been aware that when Louis died it would pass to Mary Ann's surviving children. Louis was forced into a court case which must have been most unpleasant. The court agreed that it must pass to Mary Ann's children[95] and Annie apparently told Louis that she was leaving.

Sadly, there was information on Louis Langoulant's death certificate with the record of Clementina's father taking his own life by shooting himself in the boarding house on Wellington Street here in Perth. Louis was 80 years old when he died.[96]

---

[93] *The West Australian*, 9 December 1881, page 2

[94] *The West Australian*, 24 July 1889, page 3

[95] *The West Australian*, June 1893: Langoulant v Langoulant

[96] *The West Australian*, 25 April 1894, page 3: An old man shoots himself

Louis had three children with Annie (Mary Ann III, Isaac and Leah) and when he died in 1894 they were aged 7, 5 and 3.

I am not sure of what happened to Annie, but Leah Langoulant appears to have been reared in the Girls' Orphanage on Adelaide Terrace in Perth until she 'went out into the world' to work at Pinjarra. She married Clarence Nettle when she was 29 years old and her only son was named Melville Louis George[97] after his grandfather.

Possibly Mary Ann (III) was at the Girl's Orphanage too. Isaac married Harriett Stevenson in 1926 and had three sons to help carry on the Langoulant name. He died in Geraldton aged 71.

Of Louis Langoulant's first family, there were three sons to carry on his name. John had no children. Henry had five sons and five daughters. Joe had four boys and two girls.

This second generation were all cousins to our grandmother Minnie (Clementina) McCarthy Meehan.

The Langualant Children

1 – Ann Victoria Langoulant (1842-1867)

Ann married Richard Keegan (1841-1867) in Perth in 1862 when she was 20. They had four children before she died aged 25 at Waneranooka in 1867.98. Richard deserted his four motherless children who were taken in by their Aunt Mary Ann Langoulant at Freshwater Bay (Butler's Swamp). This was in Perth and where the family home 'Pleasant Valley" had just been completed. Along with their grandparents there were 6 of their youngest children still living there. There were also two more babies born after they moved in. It must have been a crowded house since it only had five rooms.

2 – Emma Amelia Langoulant (1844 – 1912)

Emma married William Charles Vile Eaton in York in 1865. They had 7 children but only three survived infancy. Emma died in Wagin aged 68 years.

3 – Catherine Mary Langoulant (1846 – 1846)

Catherine died as a baby. Phillip may have been her twin, but I have only one reference to that name.

4 – Florentine Mary Ellen Langoulant (1848 – 1920)

Florentine was born in Perth and married Edwyn Baldwyn Powell about 1861. They had eight children, of whom 7 survived to adulthood. Florentine died in Albany aged 72.

5 – Clemens Augustus Langoulant (1850 –)

Clemens possibly died as a baby.

6 – Clementina Augusta Langoulant (1851 – 1920)

Clementina was J.P. Meehan's mother in law. She was born in Perth and married Denis Domenic McCarthy in 1869 at

---

[98] Richard had arrived in WA as a baby with his parents (Thomas and Eliza Keegan) on the *Ganges* in on 15 October 1841. Richard also had a baby brother John who was born and died soon after they arrived. He was a blacksmith and farmer at Greenough where he was doing well enough to employ 2 or 3 men. Bicentennial Dictionary: http://www.friendsofbattyelibrary.org.au/BD%20WA.htm, accessed 17Nov16.

Fremantle. At the time of her death she was 69 years of age.
They had 9 children. But Evelina died in infancy. Arthur
died in a railway accident and Ivanhoe died in action in WW1
so Percy was the sole surviving son.

7 – Maude Louisa Mary Langoulant (1853 – 1882)

Louisa was born in Perth and married Frederick John
Scanlan there in 1873 when she was 20 years old. They had
one daughter and Louisa died in Adelaide aged 29 years.

Louis was off chasing gold from before Maude was born until
1859, so Mary Ann had a break from more babies for six years.
She then had three in three years.

8 – Matilda Sophia Langoulant (1859 – 1925)

Matilda was born in Perth and married Thomas Edwin
Whitfield. There were ten children but only six appear to
have reached adulthood. She died in Perth aged 66 years.

9 – Annie Eliza Langoulant (1860 – 1860)

Annie was born in Perth but only lived for a month.

10 – John Andrew David Langoulant(1861 – 1903)

John was born in Perth and died there aged 42 years. There
is no spouse or children listed.

11 – Henry Louis Charles Langoulant (1864 – 1953)

Henry married Susan Jones at Perth in 1889 in an Anglican
ceremony. They had 11 children. He also used the name
"Charles Henry Louis Langoulant". Henry was educated by
his father and is recorded as 'Liberal' and Church of England.
He was a contractor at Claremont in 1889 and a railway
guard at Beverley in the 1890s. Later he was a railway
foreman in charge of goods traffic at Fremantle and was still
working at 80 years old in the Fremantle wool stores.

12 – Mary Ann Langoulant (1867 – 1880)

Mary Ann was born in Perth in the new house and was
named for her mother. She died aged 13 years in a fall from a
tree.

13 – Joseph William Langoulant (1869 – 1927)

Joseph was born in Perth in the new house and married
Catherine Bourke in 1892 at Fremantle. They had six
children of whom two died at two years old. He died aged 58
at East Perth.

Mary Ann Lucilla King Langualant died in 1886.

Louis Langualant then married Hannah (Annie) Rodgers who
was about 20 years old at the time. Their first baby was born
when their father was 72 years old.

14 – Mary Ann Langoulant (1887 –)

Mary Ann was named for her half-sister who died aged 13,
seven years before she was born. Mary Anne lived at least
until 1910 when she was 13.

15 – Isaac Louis Langoulant (1889 – 1960)

Isaac was born in Perth and married Harriett Alice
Stevenson in 1926. They had three sons and he died aged 71
in Geraldton.

16 – Leah Priscilla Maud Langoulant (1891 – 1984)

Leah was born in Perth and married Clarence Melville Nettle
in 1920 when she was 29 years of age. They had one son

.oOo.

Appendix 4 – Murphy Family

Our mother Catherine (Kit) Alicia Murphy was born on the 26th July 1909 at Tambellup in Western Australia.

Mother was the youngest daughter of William Murphy and Catherine Murphy (nee Meagher) who were Irish immigrants from Cashell, Tipperary Ireland. I understand that the family arrived in Western Australia either late 1908 or early 1909.

We were told that our grandmother Catherine Murphy was pregnant with our mother Kit when the family made the voyage by ship from England (Ireland) to Australia. Catherine and William already had three little girls; Nell, Bridie and Meg. The eldest was Nell who was about four at the time I think.

Patrick Murphy

William Murphy had a brother, Patrick Murphy, and a sister, Margaret Murphy, who had both moved to Western Australia quite some time before William and Catherine came here, and they also came under very different circumstances to their brother and sister in-law. Thanks to WAGS (Western Australia Genealogy Society) I have found out a little more about William and Catherine's family here in Western Australia.

From what I can remember Mother telling us, it may have been, Patrick (William's brother) who owned the farm that William and his family lived on in Quairading. Great Uncle Patrick was Mother's uncle.

Margaret Murphy

Our grandfather William Murphy and his brother Patrick had a sister called Margaret already living in Western Australia before they arrived. Margaret was married to James Joseph Haggerty who was also a farmer at Quairading.

Margaret Murphy Haggerty was our great-aunt, and James Haggerty was our great uncle by marriage.

The Butterly family

Margaret Meagher

Our grandmother, Catherine Meagher Murphy, also had a sister called Margaret who had moved to Western Australia before they did. She was Margaret Meagher, who married Burnett Butterly, a descendent of James Butterly[99] and Catherine Quinn[100].

Their descendent Burnett Arthur Butterly married Margaret Meagher (I am still searching for their marriage date and certificate). They went on to have three children who were our mother's first cousins.

Bernard Butterly

The first of the cousins was named Bernard James Lawrence. He was born on the 30th March 1915. Bernard Butterly married Thelma Curtis.

John Butterly

Their second son John was born on the 17th January 1918. John died on the 18th April 1939 in Busselton, Western Australia aged 21. This was only three years after Mother was married.

William Butterly

Margaret and Burnett's third son, William Joseph, was born on the 31st March 1920 and he died on the 13th July 1992. William Butterly was married to Gladys Mary Reynolds.

Margaret Meagher Butterly died in Busselton in 1954. Her gravestone reads 1884-1954. Her husband, Burnett Butterly,

---

[99] James Butterly had come to Western Australia in 1849 as an indentured labourer on the ship *Merope*. He worked as a farm labourer for William Chidlow of Springfield, Northam.

[100] Catherine Quinn came out to Western Australia on what was known as the "Bride Ship *Clara*" in 1853. She also went to work for the Chidlow family in Northam, where she met and married James Butterly in 1855.

was born in York, Western Australia on the 27th August 1885 and he passed away in Busselton on the 16th October 1974.

Margaret Meagher Butterly was our great aunt, and Burnett Butterly was our great uncle by marriage.

These people were very important to our mother as they were 'family'.

William and Catherine Murphy

William and Catherine Murphy both came from Cashell, which is in the County of Tipperary Ireland. Catherine Murphy's mother (our great grandmother) was Bridget O'Donnell and her father (our great grandfather) was Lawrence Meagher. I have heard it said that her family were wealthy landowners in Ireland at one time.

William Murphy and Catherine O'Donnell were married in 1904, at Cashell, County of Tipperary in Ireland. Their first three daughters – Nell (Ellen), Bridie (Bridget) and Meg (Margaret) – were all born in Ireland in Cashell. County of Tipperary.

Grandfather William Patrick Murphy and his wife (our grandmother Catherine Meagher Murphy) came to Western Australia in 1908 with their three daughters. As far as we can find out they first lived for some time in Tambellup, a small country town in Western Australia about 100km north of Albany.

Our mother Catherine (Kit) and her brother Thomas Patrick were born in Tambellup; Mother was born in 1909 and Tom was born in 1911. Our grandparents registered Mother's birth in Albany.

William Patrick Murphy died on August 6th 1925 when he was aged 55 years. William had been injured in a work related accident and it would seem that he may have died from those injuries. He was a patient at the York Hospital where he died and was then buried in the York cemetery a week or so later.

Catherine Murphy died in 1937, at Quairading Western Australia; she suffered from a heart condition and died from a heart attack.

With further searching I unearthed the following information about our grandparents. William and Catherine were on the Electoral roll for the electorate of Katanning. William was listed as a farmer of Tambellup and Catherine a housewife; that was in 1911.

This could mean that William Patrick Murphy had a farm at Tambellup, but there was also a Patrick Murphy listed as a farmer of *Calshel Park* in Tambellup. It also may be that Patrick was a brother of William's and that Patrick got William to run the farm in Tambellup for him, while he took care of his other business.

William and Catherine Murphy are not listed on the 1914 Electoral roll for Katanning.

The next time I found them on the Electoral roll was in 1916 when they gave their address then as 21 Hubble Street East Fremantle, in the electorate of North East Fremantle.

In 1925 they were shown as living in Quairading, at Jennaberring Road. William had given his occupation as a confectioner on the electoral roll that year. He may also have been farming as well, but I could not find that mentioned. I think also that William was working as a labourer while his family were living on the farm at Quairading, which is when he was badly injured and later died from those injuries at the York Hospital. The farm where the family were living was owned by William's brother Patrick as I understand from my sister Elizabeth.

While they were living in East Fremantle, our grandfather when it came time to fill out that electoral roll, wrote when he got to the part where he had to say what he did for a living, that he was a labourer.

I have found more information on William and Catherine which I will try to cover here.

As I have mentioned before, our grandfather was a Professor of languages (Gaelic) when he was living in Ireland, but here in Western Australia he worked as a farmer and a labourer as well as a confectioner. He moved the family from Tambellup to East Fremantle, then again they moved from there to Jennaberring Road in Quairading, where they all lived for about fourteen years.

Our grandfather died when our mother was still quite young; he had been ill for some time, possibly as the result of injuries caused from an accident in 1925. Mother would have been about sixteen at the time.

Our grandmother (Catherine Murphy), with the help of her daughters Bridie and Meg, opened up Tearooms in the Quairading Township. I think the name of the tearooms was Lilac. Bridie ran the tearooms, for many years, I understand.

Our mother Kitty left home to go nursing. She put her age up so that she would be accepted by the Kalgoorlie Hospital, but was found out later and had her salary docked for some time, from what she told the family.

Nell and Meg also left home. Nell worked as a barmaid in Northampton and Fremantle, I think it was. Meg left home not long after Nell and worked in different country towns till she met and married her husband, Jack Morrissey. He was a prospector and also did farm work. I also understand that Jack Morrissey was a returned soldier from the First World War.

I have listed below information on the family that I have been able to find.

Eileen (Nell) Mary Murphy Partlon

Aunty Nell (Eileen Mary, Ellen, known as Nell) Murphy, married William Patrick Partlon on the 14th February 1938. Nell died in 1970, aged 65 years. Uncle William Partlon died in August 1955 when he was 56 years old. They are both buried at the Karrakatta Cemetery in Perth Western Australia.

Nell and William Partlon had two children. The eldest was Geraldine, born in 1939. Mary Catherine (Mem) was born in 1940. Mem was named after her mother and grandmother, both of whom were Catherine.

> Our cousin Geraldine married Ken Miles and they had four children; Angela, Lucille, Roger, and Nicholas.
>
> Sadly, in 1977 Geraldine (then aged 38 years) with her children Lucille Helen aged 9 years, Roger William, aged 6 years and Nicholas Anton aged 2 years died in a car accident on a country road; Angela was the only one who survived and went on to recover from her injuries. They were all buried at the Karrakatta Cemetery in Perth W.A.
>
> Angela Miles married Lawrence Guala and they have one child Angus, born on the 25th September 2007.

Our cousin Mary Catherine Partlon (Mem) was born on the 2nd May 1940; Mary lived in Geraldton where she was born for a number of years before her death. Mem died from cancer of the oesophagus on the 31st January 2010, at the St John of God Hospice in Geraldton.

Mem's funeral was held on the 8th February 2010 at the Utakarra Crematorium Chapel in Geraldton. Rev. Jeremy Rice officiated. Tom Jones' "I'm Coming Home" was played as Mem's coffin was wheeled in.

Angela, Mem's niece, read the Eulogy. Tributes and readings were also read out by Mem's friends, Julie Burnett and Janice Bradley as well as by Jeana Sullivan. Mem's cousin Virginia O'Shannassy wrote and read a tribute to her cousin Mem on behalf of the family. Mark Knopfler's "Golden Heart" was also played. The service closed with Elvis Presley's "How Great Thou Art".

Bridget (Bridie) Francis Murphy

Our Aunt Bridie (Bridget Frances) never married and died in October 1962 aged 53. Bridie is buried at the Karrakatta Cemetery Perth.

Margaret (Meg) Murphy Morrissey

Aunty Meg (Margaret) Murphy was married to Jack Morrissey. Uncle Jack died in the 1950s and they had no children. Meg died in August 1973, aged 64. Both are buried at the Karrakatta Cemetery in Perth.

Catherine (Kit) Murphy Meehan

Our mother Catherine (Kit) Murphy married James Arthur (Arthur) Meehan in Meekatharra on the 24th October 1936. Arthur and Catherine Meehan had us six children;, Mary Philomena, Maureen Catherine, Margaret Frances Virginia (known as Virginia), Judith Anne, (Judy) Elizabeth Carmen Miranda (Liz) and John Patrick.

Thomas (Tom) Murphy

Our Uncle Tom was the youngest child and only son of William and Catherine Murphy. Thomas Murphy (our mother's brother and our uncle) lived in Northam I believe. The family think this was till his death in the 1970s. We had little to do with Uncle Thomas while we were growing up because of living in the Pilbara. I understand that Uncle Thomas was married to a lovely lady called Eileen, but we never met her. The only time I can remember meeting Uncle Tom was when our Aunty Bridie died in 1962

.oOo.

Appendix 5 – Our Irish Heritage

From what I have been able to find out, members of Grandmother's family (the Meaghers) were much involved in fighting the English; it is possible that they were members of the "Young Ireland Party". Grandfather and Grandmother Murphy were very secretive about Ireland. My mother and her sisters were not allowed to talk about the family's life before they came to Australia, so I have very little first-hand information I am afraid.

We do know that because of this involvement our grandmother's family were adversely affected; our mother told us that three members of her mother Catherine Murphy's family – two brothers and their cousin – were hanged, drawn and quartered for treason during the 1800s or earlier. From what the family have told us, our grandmother never fully coped with this tragedy.

There was another of our grandmother's relatives, Thomas Francis Meagher; and he may have been a great uncle or a cousin of grandmother's.

Thomas Francis Meagher was given the same death sentenced as other family members – that is he was sentenced to be hanged, drawn and quartered – but fortunately for him, that was commuted and instead he was sentenced to life in Tasmania as a convict.

Catherine and William found it very painful to even talk about Ireland and what had happened to their families.

Our mother and her sisters have told us that Catherine Murphy (their mother) and William (their father) became very paranoid as a result of Catherine losing her relations in such a manner. Because of their life in Ireland and the history that went with it; William and Catherine were always anxious about their own and their children's safety. Even though they were living here in Australia.

Although William Patrick Murphy was not physically involved in the "Young Ireland Party" he was vocal in his support for them and what they were trying to achieve.

This support of the "Young Ireland Party" ended up putting our grandfather in danger. Being so out-spoken about the English was in itself high risk because you never knew who would be listening. And as our grandfather at that time was a Professor of Languages at the Dublin University, he was considered to be rather foolish by his peers.

William's friends could see that by being so out-spoken it was going to end up putting William in gaol, and possibly even sentenced to death.

As things heated up in Ireland, Grandfather and his family found themselves more and more at risk, so our grandfather turned to his friends in the "Young Ireland Party " for help in leaving Ireland.

Members of the "Young Ireland Party" managed to arrange a berth on a ship heading for England and then Australia for Grandfather and Catherine and their three little girls who would have been about four, three and two at that time.

I understood from our mother that there is the possibility that Grandfather and Grandmother Murphy may have been forced to change their name before leaving Ireland, but I have not been able to establish if that was the case.

Because of what we do know and because of the risk that our grandparents and their friends from the "Young Ireland Party" faced if they were caught helping our grandparents, there is a strong possibility this did happen. So we may never know grandfather's back ground or even what his history was before they settled in Australia. What we do know was that William Patrick Murphy was haunted by his past.

From the stories that our mother and aunts told us when we were growing up, Grandfather was a very troubled man who was burdened by fear and guilt at leaving his friends behind in Ireland.

William Patrick Murphy, it would appear, spent his life here in Australia afraid of being hunted down and forced to return to Ireland where he could possibly have met the same fate that other Irish men before him had met.

I can remember after I had grown up my mother joking about her convict back ground. She would say if you looked close enough you would see the marks of the chains around her ankles. She also talked about her uncles being convicts; she was always very proud about that. Mother told me that I must never forget where I came from and the history that went with it. At that time in my life I never really took a lot of notice which I now regret.

This chapter is taken from what I can remember my family telling me. I have no documentation just hearsay as they have all passed on. Aunty Nell and Aunty Meg also spoke about what they knew or could remember of their time in Ireland.

Grandfather and his family arrived in Western Australia in either 1908 or 1909. It would appear that our grandfather then set about building a new life for himself and his family.

There was to be no continuation of Grandfather's University career here. Once they arrived in Western Australia, Grandfather went about making sure that he and his family blended in with the local population. The type of work that he did here was always low key. The family were also expected to blend in and not stand out from what I have been able to find out. Grandfather was always very nervous and moved the family a lot. The children were trained not to say anything about the family.

William it would appear was a very tortured soul and his family paid the price. It must have been very hard for him to go from being an academic in one life, to becoming a farmer and labourer here in Western Australia. William, as I understand, worked very hard at keeping a low profile after the family moved from Ireland.

I often wondered if our grandfather William Patrick Murphy may have turned into an alcoholic after they moved here. No one ever said so, but after all they had been through, it has left me wondering. I would not have blamed him if he had been. Life was very hard for them back then I think.

.oOo.

Thomas Francis Meagher

There has been much written about our famous family member, Thomas Francis Meagher[101], who was deported to Tasmania Australia as an Irish convict in 1849 when he was about twenty six years of age. He was a relative of my Grandmother Catherine Meagher Murphy – possibly her father's cousin or uncle.

Our cousin Mary Partlon could remember our aunties, Meg Morrissey and Bridie Murphy, as well as her mother, Nell Partlon and our mother Catherine (Kit) Meehan in later years, when we had all grown up, talking in hushed tones about their mother, our grandmother Catherine Meagher Murphy, having lost members of her family during the Irish Troubles as they called that time.

My cousin Mem (Mary) Partlon was able to find information about our distant cousin Thomas Francis Meagher while on a visit to Tasmania a few years back. Mem rang me very excited to say that she had found the following information, which I have included below.

Thomas Francis's father was the Lord Mayor of the city of Waterford in Ireland. Thomas was born in Waterford, Ireland on the 3rd August 1823 and at the time of his death in 1867 was only 44 years of age.

By the time Thomas was in his early twenties he had built a reputation for himself as an orator and journalist of some repute. He was given the nickname of "Meagher of the Sword".

Before Thomas went to America, he had led a very interesting if somewhat precarious life back in Ireland. Thomas had taken on a rather prominent role in his country's politics as one of the leaders of the "Young Ireland Party".

---

[101] Courtesy: of the Hobart Library, in Tasmania.

Thomas was arrested and became a political prisoner. He was sent to trial where he was tried and convicted on the charge of treason; he was then sentenced to be hanged, drawn and quartered. This sentence however was eventually commuted to being banishment for life to Van Diemen's land (Tasmania, just off the mainland of Australia). Thomas was there from 1849 to1852. While a political prisoner, he took the surname O'Meagher[102] by his choice.

As a political prisoner, Thomas was able to give his parole (promise not to escape) and was fairly free to live in the community. He met and married Katherine Bennett[103], although his fellow Irish political prisoners strongly advised against it. She was a free-born governess when he met her but her father had been a convict while her mother had come free from Ireland to join her husband.

After three years, an escape was arranged, and Thomas informed the authorities he was handing back his parole and promptly sped away to the waiting ship. This was in 1852, and he left behind his wife Katherine, who was almost due to have their first child.

Sadly their son died at four months from influenza. He was buried at St John's Catholic Church at Richmond in Tasmania. As a tribute to his famous father, the grave was later moved to beside the front door of the church. With the help of friends Thomas was able to obtained passage on a sailing vessel which was heading to America. Katherine and their baby were intended to follow Thomas to America after the baby's birth.

Upon his arrival there he was greeted with great enthusiasm by the Americans and especially the Irish Americans. Thomas was

---

[102] Wikipedia, the Free Encyclopaedia.
https://en.wikipedia.org/wiki/Thomas_Francis_Meagher

[103] Macfie, Peter: Thomas Meagher, Sergeant Daniel Murphy and the Bennett Family, 2011

feted with public receptions being held in his honour around the country. For many years the anniversary of his arrival was celebrated by a club named in his honour. Thomas's story was well known among the Irish Americans and his escape from death back in Ireland by being hanged, drawn and quartered was well known.

In Historic Memory of
Henry Emmet Fitzgerald O'Meagher
Infant Son of
Thomas Francis O'Meagher
And Catherine O'Meagher
Died 8th June 1852
Aged 4 months
"Suffer Little Children to Come Unto Me" Matt XIX:14

In 1853, Katherine was escorted to Ireland by Bishop Willson, and then by her father-in-law to America. When they finally caught up with Thomas, he had lost interest in her, (or so he said) and Katherine was sent packing back to his family in Ireland after about 4 months.

Figure 41: Baby Henry O'Meagher's grave (Photo Michael Horton)

Katherine was again pregnant with Thomas's child. She stayed in Ireland with her in-laws, including Thomas's father, till the birth of her son. Tragically, it was a very difficult birth and Katherine died in May 1854; she was only 22 years old. Katherine's son (Thomas Francis Meagher II after his father) was raised by family members in Ireland. He never met his father, who was apparently not interested in this child. However, they did apparently exchange letters occasionally.

Thomas travelled around America on lecture tours till finally he settled in California in 1854. There he went on to read and practice law, and he also edited a weekly paper, the "Irish News"

In 1856, Thomas visited Nicaragua and Costa Rica. I am not sure why he went there but think it might have something to do with his law practice or the weekly paper that he was editing.

That same year he married again, to Elizabeth Townsend of New York, but had no further children.

He fought in the American Civil War on the side of the Union, and went on to be appointed Secretary of the new Territory of Montana where he was also Acting Governor at the time of his death.

Thomas Francis Meagher fell overboard from a river boat into the Missouri River at Fort Benton on 1st July 1867.[104] (There was much discussion about whether he fell or was pushed.)

A dispatch was sent to the New York Times, on 8th July 1867, informing them of the death by drowning of General Thomas Francis Meagher.[105]

---

[104] Wikipedia, the Free Encyclopaedia. https://en.wikipedia.org/wiki/Thomas_Francis_Meagher

[105] As apparently had already been arranged, the widowed Mrs Elizabeth Meagher took charge of her stepson from Ireland. He lived with her in New York from the early 1870s and entered West Point Military Academy in 1872 (when he was 18 years old). This did not last, and he also fell out with his stepmother. Thomas II went on to marry Mary Lavinia Carpenter of Sacramento and lived the life of a gentleman in San Francisco. He later worked in Manilla (Phillipines) where he died in 1910, leaving one son (another Thomas Francis Meagher) and a daughter.

This is a poem that Thomas Francis Meagher wrote sometime later about his time in Tasmania.

"I WOULD NOT DIE"

I would not die in this bright hour,
While hope's sweet dream is flowing:
I would not die while Youth's gay flower
In springtide pride is glowing.

The path I trace in fiery dreams
For Manhood's flight, to-morrow
Oh, let me tread 'mid those bright gleams
Which souls from Fame will borrow.

I would not die! I would not die!
In Youths 'bright hour of pleasure;
I would not leave, without a sigh,
The dreams, the hopes, I treasure!

I set young seeds in earth to-day,
While yet the sun was gushing;
And shall I pass, ere these, away,
Nor see the flowerets blushing?

Are those young seeds, when earth looks fair,
To rise with fragrance teeming,
And shall the hand that placed them there
Lie cold when they are gleaming?

I would not die! I would not die!
In Youth's bright hour of pleasure;
I would not leave, without a sigh,
The dreams, the hopes, I treasure!

Written by Thomas Francis Meagher.[106]

.oOo.

---

[106] From "Shadow over Tasmania – the whole story of the convicts" by Coultman Smith, 1955.Published by The University of California. The information was located courtesy of the Hobart Library in Tasmania, Australia

Table of Figures

All photos are the property of Virginia Meehan O'Shannassy unless otherwise indicated. Front and back cover photos – Bree Krieger.

Index

290

* 9 7 8 0 6 4 6 9 6 4 8 3 6 *